Hacking the Kinect

Jeff Kramer
Nicolas Burrus
Florian Echtler
Daniel Herrera C.
Matt Parker

Hacking the Kinect

ISBN-13 (pbk): 978-1-4302-3867-6

ISBN-13 (electronic): 978-1-4302-3868-3

President and Publisher: Paul Manning
Lead Editor: Jonathan Gennick
Technical Reviewer: Curiomotion LLC
Editorial Board: Steve Anglin, Mark Beckner, Ewan Buckingham, Gary Cornell, Louise Corrigan, Morgan Ertel, Jonathan Gennick, Jonathan Hassell, Robert Hutchinson, Michelle Lowman, James Markham, Matthew Moodie, Jeff Olson, Jeffrey Pepper, Douglas Pundick, Ben Renow-Clarke, Dominic Shakeshaft, Gwenan Spearing, Matt Wade, Tom Welsh
Coordinating Editor: Anita Castro
Copy Editor: Heather Lang
Compositor: Bytheway Publishing Services
Indexer: SPI Global
Artist: SPI Global
Cover Designer: Anna Ishchenko

Distributed to the book trade worldwide by Springer Science+Business Media New York, 233 Spring Street, 6th Floor, New York, NY 10013. Phone 1-800-SPRINGER, fax (201) 348-4505, e-mail orders-ny@springer-sbm.com, or visit www.springeronline.com.

For information on translations, please e-mail rights@apress.com, or visit www.apress.com.

Apress and friends of ED books may be purchased in bulk for academic, corporate, or promotional use. eBook versions and licenses are also available for most titles. For more information, reference our Special Bulk Sales–eBook Licensing web page at www.apress.com/bulk-sales.

Any source code or other supplementary materials referenced by the author in this text is available to readers at www.apress.com. For detailed information about how to locate your book's source code, go to www.apress.com/source-code/.

I dedicate this book to Jennifer Marriott, the light of my life.

—Jeff Kramer

For my family and for Lauren, who helped me in every possible way.

—Matt Parker

Contents at a Glance

About the Authors x

About the Technical Reviewer xiii

Acknowledgments xiv

Chapter 1: Introducing the Kinect 1

Chapter 2: Hardware 11

Chapter 3: Software 41

Chapter 4: Computer Vision 65

Chapter 5: Gesture Recognition 89

Chapter 6: Voxelization 103

Chapter 7: Point Clouds, Part 1 127

Chapter 8: Point Clouds, Part 2 151

Chapter 9: Object Modeling and Detection 173

Chapter 10: Multiple Kinects 207

Index 247

Contents

▨ **About the Authors**.. **x**

▨ **About the Technical Reviewer** .... **xiii**

▨ **Acknowledgments** .... **xiv**

▨ **Chapter 1: Introducing the Kinect**.. **1**

Hardware Requirements and Overview 1

Installing Drivers.. 2

    Windows 2

    Linux 6

    Mac OS X 8

    Testing Your Installation 9

    Getting Help 9

Summary 9

▨ **Chapter 2: Hardware**... **11**

Depth Sensing 11

RGB Camera... 13

Kinect RGB Demo.. 13

    Installation 14

    Making a Calibration Target 16

    Calibrating with RGB Demo 17

Tilting Head and Accelerometer 23

Summary 39

■ **Chapter 3: Software** **41**

Exploring the Kinect Drivers 41

OpenNI 41

Microsoft Kinect SDK 41

OpenKinect 41

Installing OpenCV 42

Windows 42

Linux 44

Mac OS X 46

Installing the Point Cloud Library (PCL) 48

Windows 48

Linux 52

Mac OS X 53

Summary 63

■ **Chapter 4: Computer Vision** **65**

Anatomy of an Image 65

Image Processing Basics 66

Simplifying Data 66

Noise and Blurring 67

Contriving Your Situation 69

Brightness Thresholding 69

Brightest Pixel Tracking 72

Comparing Images 74

Thresholding with a Tolerance 75

Background Subtraction 76

Frame Differencing 80

Combining Frame Differencing with Background Subtraction 85

Summary 87

▓ **Chapter 5: Gesture Recognition** **89**

What Is a Gesture? 89

Multitouch Detection 89

    Acquiring the Camera Image, Storing the Background, and Subtracting 92

    Applying the Threshold Filter 92

    Identifying Connected Components 93

    Assigning and Tracking Component IDs 95

    Calculating Gestures 96

    Creating a *Minority Report*—Style Interface 99

    Considering Shape Gestures 101

Summary 101

▓ **Chapter 6: Voxelization** **103**

What Is a Voxel? 103

Why Voxelize Data? 104

Voxelizing Data 105

Manipulating Voxels 107

Clustering Voxels 120

Tracking People and Fitting a Rectangular Prism 122

Summary 125

▓ **Chapter 7: Point Clouds, Part 1** **127**

Representing Data in 3-D 127

    Voxels 128

    Mesh Models 128

    Point Clouds 128

Creating a Point Cloud with PCL 129

Moving From Depth Map to Point Cloud 130

Coloring a Point Cloud 131

From Depth to Color Reference Frame 131

Projecting onto the Color Image Plane 132

Visualizing a Point Cloud 132

Visualizing with PCL 133

Visualizing with OpenGL 133

Summary 150

■ **Chapter 8: Point Clouds, Part 2** **151**

Registration 151

2-D Registration 152

3-D Registration 155

Robustness to Outliers 157

Simultaneous Localization and Mapping (SLAM) 159

SLAM Using a Conventional Camera 159

Advantages of Using the Kinect for SLAM 160

A SLAM Algorithm Using the Kinect 160

Real-Time Considerations 161

Surface Reconstruction 162

Normal Estimation 162

Triangulation of Points 162

Summary 172

■ **Chapter 9: Object Modeling and Detection** **173**

Acquiring an Object Model Using a Single Kinect Image 173

Tabletop Object Detector 174

Fitting a Parametric Model to a Point Cloud 178

Building a 3-DModel by Extrusion 179

viii

Acquiring a 3-D Object Model Using Multiple Views 188

Overview of a Marker-Based Scanner... 189

Building a Support with Markers 191

Estimating the 3-DCenter of the Markers in the Camera Space.................................. 191

Kinect Pose Estimation from Markers 192

Cleaning and Cropping the Partial Views 195

Merging the Point Clouds 196

Getting a Better Resolution... 199

Detecting Acquired Objects 200

Detection Using Global Descriptors 200

Estimating the Pose of a Recognized Model 204

Summary 206

Chapter 10: Multiple Kinects 207

Why Multiple Kinects? 207

The Kinect Has a Limited Field of View 207

The Kinect Fills Data from a Single Direction Only 207

The Kinect Casts Depth Shadows in Occlusions 208

What Are the Issues with Multiple Kinects? 208

Hardware Requirements.. 209

Interference Between Kinects 209

Calibration Between Kinects 209

Interference 209

Calibration 228

Summary 246

Index 247

About the Authors

 Jeff Kramer (@Qworg) is a builder, maker, hacker, and dreamer. Jeff is currently a research programmer and roboticist at the National Robotics Engineering Center (NREC) in Pittsburgh, PA (www.rec.ri.cmu.edu/). He also breaks things and rebuilds them all of the time. He's taught graduate courses in animal psychology, chaired international robotics conference sessions, made Bill Nye cry with a weapon of mass destruction, constructed high-powered arc lamps, written journal articles in robotics, devastated players with a robotic foozball table, and generally made a mess—a beautiful, strangely functional mess that is always evolving. Jeff holds a master's degree in robotics from the University of South Florida. You can check him out at http://mind-melt.com/ and http://about.me/JeffreyKramer/. Or just e-mail him at jeffkramr@gmail.com.

◾ Nicolas Burrus is a researcher in computer vision at the Carlos III University of Madrid, with a special interest in 3D object model acquisition and recognition for robotic applications. He actively took part in the impressive wave of interest that followed the release of the Kinect by publishing RGBDemo, an open source software showcasing many applications of the Kinect. RGBDemo is being used by many research labs, companies, and hobbyists, and this success led him to co-found the Manctl startup with the ambition of developing a low-cost universal 3D scanner. He holds a PhD from Paris VI University and a master's of computer science from EPITA (Paris, France).

◾ Florian Echtler is a researcher in human-computer interaction and information security, currently working at Siemens Corporate Technology in Munich, Germany. He wrote the very first Kinect hack, an interface in the style of *Minority Report,* in less than 24 hours after the first release of the open source drivers. He is the main author of libTISCH, a NUI development platform that supports the Kinect as an input device and is actively used in several research projects. He holds a diploma and a PhD in computer science from the Technical University of Munich (TUM).

Daniel Herrera C. works as a computer vision researcher at the University of Oulu, Finland. He is currently doing his PhD in the areas of image-based rendering and free viewpoint video. His early work with the Kinect led him to develop an open source calibration algorithm. The Kinect Calibration Toolbox is now being used by researchers around the world. Since then the Kinect has become one of the main tools for his research. He holds a master's in computer vision and robotics from the Erasmus Mundus program Vibot (UK, Spain, and France).

Matt Parker is a new media artist and game designer. As an artist, his interest lies in exploring the intersection of the physical and digital worlds. His work has been displayed at the American Museum of Natural History, SIGGRAPH Asia, the NY Hall of Science, Museum of the Moving Image, FILE Games Rio, Sony Wonder Technology Lab, and many other venues. His game Lucid was a finalist in Android's Developer Challenge 2, and his game Recurse was a finalist for Indiecade 2010. His project Lumarca won the "Create the Future" Prize at the World Maker Faire 2010.

Matt earned his BS in computer science from Vassar College and a master's from NYU's Interactive Telecommunications Program (ITP). He has served as a new media researcher and adjunct faculty member at NYU since 2009. He is currently a visiting professor at Sarah Lawrence College and an artist in residence at Eyebeam Art and Technology Center.

About the Technical Reviewer

Curiomotion LLC is a technology company dedicated not to providing a specific product or service, but to bringing tomorrow's technology to today's applications. Currently, Curiomotion's team of software engineers is working with the latest advancements in motion sensing technology to enhance consumer engagement in retail scenarios. Curiomotion is one of the first companies to offer products for this new paradigm of technology-driven commerce, providing interactive marketing solutions compatible with any business strategy.

Acknowledgments

A book of this magnitude cannot be accomplished alone. First, I would like to thank everyone at Apress for giving me the opportunity to write about something truly exciting and game changing. I'm proud that we are supporting future creative and commercial endeavors using the Kinect and other 3D sensors. I would like to thank Jonathan Gennick, my lead editor, who first reached out to me and offered just a chapter, then half a book, and then a book. Without him, I would have never started. I would like to thank Anita Castro, my coordinating editor, who dealt with my fuzzy deadlines, last-minute changes, and never-ending shenanigans with grace and dignity. I know I put her through hell, and I simply cannot apologize enough. Without her, I would have never finished. Technical reviewer Max Choi's insight not only caught bugs and hammered on inconsistencies, but also truly turned the examples from mere toys to finished products. Copy editor Heather Lang's deft touch polished my oft-times clunky prose and refitted all of the text—a monumental task. Thank you both. Michelle Lowman, part of the editorial board, also deserves thanks for her role in refining the ideas that led to this book.

Working with genuinely amazing people never gets old. I would like to thank Ed Paradis (http://edparadis.com/) for his unending support, as well as his deft photography and Lego skills, especially on the final chapter. He was an excellent person to bounce ideas off of and helped me get my chapters finished. I would like to thank my coauthors—Daniel, Florian, Matt, and Nicolas. You guys took a chance on me, on this book, and it panned out! Like jumping into a very cold lake—you just have to start swimming. Thank you for your efforts and contributions. This book would not have been even a quarter as good without you.

I would like to thank Kyle Wiens over at iFixit (www.ifixit.com) for letting me use one of their teardown images, and for providing such an amazing resource for hardware hackers everywhere. If you can't fix it, you don't own it. I would like to thank the team building PCL (http://pointclouds.org/) for all of their crucial work—they are truly bringing 3D manipulation to the masses.

I would like to thank Dr. Abraham Kandel. As always, your support from afar carries me through rough spots, and your example inspires me to always do more, better. Thank you.

I would like to thank my mom and dad for providing such an excellent example and their support growing up. I would have never been in this position without them.

Last and most certainly not least, I would like to thank my soon-to-be wife, Jennifer Marriott, and our lovely daughter, Charlotte. Late nights, missed dinners, angst, misery, and snark—you both stood by me, supported me, and loved me, even when it was hard to do so. I love you both in ways I cannot express in words. Thank you!

Jeff Kramer

I wish to thank all those who have called me their friend in this cold city, for any endeavor is rendered empty if there is nobody to share it with. And a special mention to those who have fed me, for I often forget to do it myself.

Daniel Herrera C.

I'd like to thank NYU ITP, Eyebeam Art and Technology Center, and the Openframeworks community.

Matt Parker

CHAPTER 1

Introducing the Kinect

Welcome to *Hacking the Kinect*. This book will introduce you to the Kinect hardware and help you master using the device in your own programs. We're going to be covering a large amount of ground—everything you'll need to get a 3-D application running—with an eye toward killer algorithms, with no unusable filler.

Each chapter will introduce more information about the Kinect itself or about the methods to work with the data. The data methods will be stretched across two chapters: the first introduces the concept and giving a basic demonstration of algorithms and use, and the second goes into more depth. In that second chapter, we will show how to avoid or ameliorate common issues, as well as discuss more advanced algorithms. All chapters, barring this one, will contain a project—some basic, some advanced.

We expect that you will be able to finish each chapter and immediately apply the concepts into a project of you own; there is plenty of room for ingenuity with the first commercial depth sensor and camera!

Hardware Requirements and Overview

The Kinect requires the following computer hardware to function correctly. We'll cover the requirements more in depth in Chapter 3, but these are the basic requirements:

- A computer with at least one, mostly free, USB 2.0 hub.

    - The Kinect takes about 70% of a single hub (not port!) to transmit its data.

    - Most systems can achieve this easily, but some palmtops and laptops cannot. To be certain, flip to Chapter 2, where we give you a quick guide on how to find out.

- A graphics card capable of handling OpenGL. Most modern computers that have at least an onboard graphics processor can accomplish this.

- A machine that can handle 20 MB/second of data (multiplied by the number of Kinects you're using). Modern computers should be able to handle this easily, but some netbooks will have trouble.

- A Kinect sensor power supply if your Kinect came with your Xbox 360 console rather than standalone.

Figure 1-1 shows the Kinect itself. The callouts in the figure identify the major hardware components of the device. You get two cameras: one infrared and one for standard, visible light. There is an infrared emitter to provide structured light that the infrared camera uses to calculate the depth image. The status light is completely user controlled, but it will tell you when the device is plugged into the USB (but not necessarily powered!) by flashing green.

Figure 1-1. Kinect hardware at a glance

Installing Drivers

This book focuses on the OpenKinect driver – a totally open source, low level driver for the Kinect. There are a few other options (OpenNI and the Kinect for Windows SDK), but for reasons to be further discussed in Chapter 3, we'll be using OpenKinect. In short, OpenKinect is totally open source, user supported and low level, therefore extremely fast. The examples in this book will be written in C/C++, but you can use your favorite programming language; the concepts will definitely carry over.

▪ **Note** Installation instructions are split into three parts, one for each available OS to install to. Please skip to the section for the OS that you're using.

Windows

While installing and building OpenKinect drivers from source is fairly straightforward, it can be complicated for first timers. These steps will take you through how to install on Windows 7 (and should also work for earlier versions of Windows).

1. Download and install Git (http://git-scm.com). Be sure to select "Run git from the Windows Command Prompt" and "Check out Windows style, commit Unix-style line endings".

2. Open your command prompt; go to the directory where you want your source folder to be installed, and clone/branch as in Listing 1-1. See the "Git Basics" sidebar for more information.

Listing 1-1. Git Commands for Pulling the Source Code

```
C:\> mkdir libfreenect
C:\> cd libfreenect
C:\libfreenect> git clone https://github.com/OpenKinect/libfreenect.git (This will clone into
a new libfreenect directory)
C:\libfreenect> cd libfreenect
C:\libfreenect\libfreenect> git branch -track unstable origin/unstable
```

3. There are three major dependencies that must be installed for libfreenect to
    function: libusb-win32, pthreads-win32, and GLUT. Some of the options you
    select in the next section are dependent on your choice of compiler.

    a. Download libusb-win32 from http://sourceforge.net/projects/libusb-win32/.

    b. Extract and move the resulting folder into /libfreenect.

    c. Download pthreads-win32 from http://sourceware.org/pthreads-win32/.
        Find the most recent candidate with release.exe at the end.

    d. Extract and store the folder in /libfreenect. If you're using Microsoft Visual
        Studio 2010, copy /Pre-built.2/lib/pthreadVC2.dll to /Windows/System32/. If
        using MinGW, copy /Pre-built.2/lib/pthreadGC2.dll to /Windows/System32/
        instead.

    e. Download GLUT from http://www.xmission.com/~nate/glut.html. Find the
        most recent release ending in "-bin.zip".

    f. Extract and store the resulting folder in /libfreenect.

    g. Copy glut32.dll to /Windows/System32/. If you're using Microsoft Visual
        Studio 2010, copy glut.h to the /include/GL folder in your Visual Studio tree
        and glut32.lib library to /lib in the same tree. If the GL folder does not exist,
        create it. However, if you're using MinGW, copy glut.h to /include/GL folder
        in the MinGW root directory.

4. All of the dependencies are in place! Now we can install the low-level Kinect
    device driver.

    a. Plug in your Kinect. After a quick search for drivers, your system should
        complain that it cannot find the correct drivers, and the LED on the Kinect
        itself will not light. This is normal.

    b. Open Device Manager. Start ▸ Control Panel ▸ Hardware and Sound ▸ Device
        Manager.

    c. Double-click Xbox NUI Motor. Click Update Driver in the new window that
        appears.

    d. Select "Browse my computer for driver software", and browse to
        /libfreenect/platform/inf/xbox nui motor/.

    e. After installation, the LED on the Kinect should be blinking green. Repeat steps
        3 and 4 for Xbox NUI Camera and Xbox NUI Audio.

5. Download CMake from www.cmake.org/cmake/resources/software.html. Get the most recent .exe installer, and install it.

6. Make sure you have a working C compiler, either MinGW or Visual Studio 2010.

7. Launch CMake-GUI, select /libfreenect as the source folder, select an output folder, and click the Grouped and Advanced check boxes to show more options.

8. Click Configure. You see quite a few errors. This is normal! Make sure that CMake matches closely to Figure 1-2. At the time of this writing, Fakenect is not working on Windows, so uncheck its box.

■ **Note** MinGW is a minimal development environment for Windows that requires no external third-party runtime DLLs. It is a completely open source option to develop native Windows applications. You can find out more about it at www.mingw.org.

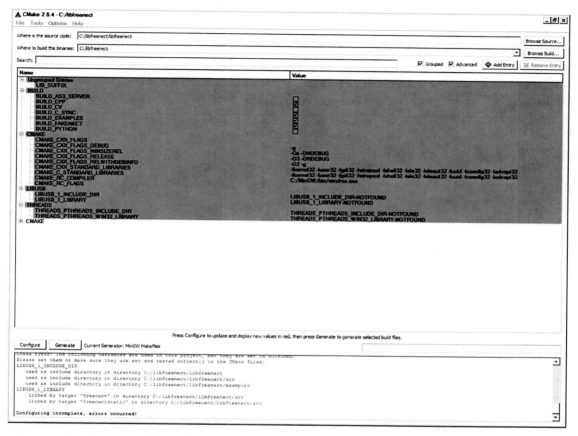

Figure 1-2. CMake preconfiguration

9. Here too, the following steps split based on compiler choices; this installation step is summarized in Table 1-1.

   a. For Microsoft Visual Studio 2010, GLUT_INCLUDE_DIR is the /include directory in your Visual Studio tree. GLUT_glut_LIBRARY is the actual full path to glut32.lib in your Visual Studio tree. LIBUSB_1_LIBRARY is /lib/msvc/libusb.lib in the libusb installation directory. THREADS_PTHREADS_WIN32_LIBRARY is /Prebuilt.2/lib/pthreadVC2.lib in the pthreads installation directory.

   b. For MinGW, the following choices must be set: GLUT_INCLUDE_DIR is the GLUT root directory. GLUT_glut_LIBRARY is the actual full path to glut32.lib in the GLUT root directory. LIBUSB_1_LIBRARY is /lib/gcc/libusb.a in the libusb installation directory. THREADS_PTHREADS_WIN32_LIBRARY is /Prebuilt.2/lib/pthreadGC2.a in the pthreads installation directory.

c. For both, the following choices must be set: LIBUSB_1_INCLUDE_DIR is /include in the libusb installation directory. THREADS_PTHREADS_INCLUDE_DIR is /Pre-built.2/include in the pthreads installation directory.

Table 1-1. CMake Settings for Microsoft Visual Studio 2010 and MinGW

CMake Setting	Microsoft Visual Studio 2010	MinGW
GLUT_INCLUDE_DIR	<MSVSRoot>/VC/include	<GLUTRoot>/
GLUT_glut_LIBRARY	<MSVSRoot>/VC/lib/glut32.lib	<GLUTRoot>/glut32.lib
LIBUSB_1_INCLUDE_DIR	<LIBUSBRoot>/include	<LIBUSBRoot>/include
LIBUSB_1_LIBRARY	<LIBUSBRoot>/lib/msvc/libusb.lib	<LIBUSBRoot>/lib/gcc/libusb.a
THREADS_PTHREADS_INCLUDE_DIR	<PTHREADRoot>/Pre-built.2/include	<PTHREADRoot>/Pre-built.2/include
THREADS_PTHREADS_WIN32_LIBRARY	<PTHREADRoot>/Pre-built.2/lib/pthreadVC2.lib	<PTHREADRoot>/Pre-built.2/lib/pthreadGC2.a

10. Dependencies that have yet to be resolved are in red. Click Configure again to see if everything gets fixed.

11. As soon as everything is clear, click Generate.

12. Open your chosen output folder, and compile using your compiler.

13. Test by running /bin/glview.exe.

■ **Note** If you have problems compiling in Windows, check out the fixes in Chapter 3 to get your Kinect running.

Linux

Installing on Linux is a far simpler than on Windows. We'll go over both Ubuntu and Red Hat/Fedora. For both systems, you need to install the following dependencies; the first line in each of the listings below takes care of this step for you:

- git-core
- cmake
- libglut3-dev
- pkg-config
- build-essential

- libxmu-dev
- libxi-dev
- libusb-1.0.0-dev

Ubuntu

Run the commands in Listing 1-2. Follow up by making a file named 51-kinect.rules in /etc/udev/rules.d/, as shown in Listing 1-3, and 66-kinect.rules in the same location, as shown in Listing 1-4.

Listing 1-2. Ubuntu Kinect Installation Commands

```
sudo apt-get install git-core cmake libglut3-dev pkg-config build-essential libxmu-dev libxi-
dev libusb-1.0-0-dev
git clone https://github.com/OpenKinect/libfreenect.git
cd libfreenect
mkdir build
cd build
cmake ..
make
sudo make install
sudo ldconfig /usr/local/lib64/
sudo adduser <SystemUserName> video
sudo glview
```

Listing 1-3. 51-kinect.rules

```
# ATTR{product}=="Xbox NUI Motor"
SUBSYSTEM=="usb", ATTR{idVendor}=="045e", ATTR{idProduct}=="02b0", MODE="0666"
# ATTR{product}=="Xbox NUI Audio"
SUBSYSTEM=="usb", ATTR{idVendor}=="045e", ATTR{idProduct}=="02ad", MODE="0666"
# ATTR{product}=="Xbox NUI Camera"
SUBSYSTEM=="usb", ATTR{idVendor}=="045e", ATTR{idProduct}=="02ae", MODE="0666"
```

Listing 1-4. 66-kinect.rules

```
#Rules for Kinect
SYSFS{idVendor}=="045e", SYSFS{idProduct}=="02ae", MODE="0660",GROUP="video"
SYSFS{idVendor}=="045e", SYSFS{idProduct}=="02ad", MODE="0660",GROUP="video"
SYSFS{idVendor}=="045e", SYSFS{idProduct}=="02b0", MODE="0660",GROUP="video"
#End
```

Red Hat / Fedora

Use Listing 1-5 to install, and then make the files in Listings 1-3 and 1-4 in /etc/udev/rules.d/.

Listing 1-5. Red Hat/Fedora Kinect Installation Commands

```
yum install git cmake gcc gcc-c++ libusb1 libusb1-devel libXi libXi-devel libXmu libXmu-devel
freeglut freeglut-devel
git clone https://github.com/OpenKinect/libfreenect.git
cd libfreenect
mkdir build
cd build
cmake ..
make
sudo make install
sudo ldconfig /usr/local/lib64/
sudo adduser <SystemUserName> video
sudo glview
```

Mac OS X

There are several package installers for OS X, but we'll be focusing on MacPorts (Fink and Homebrew are not as well supported and are too new for most users). The maintainers of OpenKinect have issued a special port of libusb-devel that is specifically patched to work for the Kinect. Move to a working directory, and then issue the commands in Listing 1-6. If you want to build an XCode project instead of a CMake one, change the cmake line to cmake –G Xcode . . . . In cmake, configure and generate before exiting to run the remaining code.

■ **Note** MacPorts is an open source system to compile, install, and upgrade software on your OS X machine. It is the Mac equivalent of apt-get. Although it is almost always properly compiles new libraries on your system, it is extremely slow due to its "reinstall everything to verify" policy. Homebrew is up and coming and will likely be the package manager of the future. To learn more about MacPorts, please visit www.macports.org.

Listing 1-6. Mac OS X Kinect Installation Commands

```
sudo port install git-core
sudo port install libtool
sudo port install libusb-devel
sudo port install cmake
git clone https://github.com/OpenKinect/libfreenect.git
cd libfreenect/
mkdir build
cd build
ccmake ..
Run Configure to generate the initial build description.
Double check the settings in the configuration (this shouldn't be an issue with a standard
installation)
Run Generate and Exit
```

```
make
sudo make install
```

Testing Your Installation

Our driver of choice, libfreenect, helpfully ships with a small set of demonstration programs. You can find these in the /bin directory of your build directory. The most demonstrative of these is glview; it shows an attempt at fitting the color camera to the 3D space. Your glview output should look much like Figure 1-3.

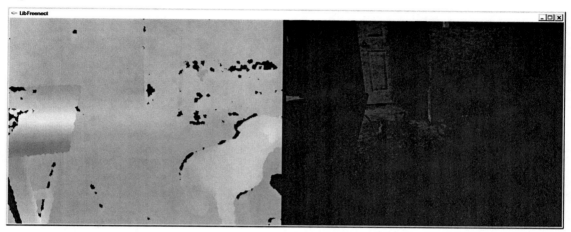

Figure 1-3. glview capture

Getting Help

While much of the information in this chapter should be straightforward, there are sometimes hiccups in the process. In that case, there are several places to seek help. OpenKinect has one of the friendliest communities out there, so do not hesitate to ask questions.

- http://openkinect.org: This is the home page for the OpenKinect community; it also has the wiki and is full of great information.

- http://groups.google.com/group/openkinect: This is the OpenKinect user group mailing list, which hosts discussions and answers questions.

- IRC: #OpenKinect on irc.freenode.net

Summary

In this first chapter, you installed the initial driver software, OpenKinect, and ran your first 3-D application on your computer. Congratulations on entering a new world! In the next chapter, we're going to dive deep into the hardware of the Kinect itself, discussing how the depth image is generated and some of the limitations of your device.

CHAPTER 2

Hardware

In this chapter, you will extensively explore the Kinect hardware, covering all aspects of the system including foibles and limitations. This will include the following:

- How the depth sensing works
- Why you can't use your Kinect outside
- System requirements and limitations

Let's get started.

Figure 2-1. Kinect external diagram

Depth Sensing

Figure 2-1 will serve as your guidebook to the Kinect hardware. Let's start with the depth sensing system. It consists of two parts: the IR laser emitter and the IR camera. The IR laser emitter creates a known noisy pattern of structured IR light at 830 nm. The output of the emitter is shown in Figure 2-2. Notice the nine brighter dots in the pattern? Those are caused by the imperfect filtering of light to create the pattern. Prime Sense Ltd., the company that worked with Microsoft to develop the Kinect, has a patent (US20100118123) on this process, as filters to create light like this usually end up with one extremely bright dot in the center instead of several moderately bright dots. The change from a single bright dot to several moderately bright dots is definitely a big advancement because it allows for the use of a higher powered laser.

Figure 2-2. Structured light pattern from the IR emitter

The depth sensing works on a principle of **structured light**. There's a known pseudorandom pattern of dots being pushed out from the camera. These dots are recorded by the IR camera and then compared to the known pattern. Any disturbances are known to be variations in the surface and can be detected as closer or further away. This approach creates three problems, all derived from a central requirement: light matters.

- The wavelength must be constant.

- Ambient light can cause issues.

- Distance is limited by the emitter strength.

The wavelength consistency is mostly handled for you. Within the sensor, there is a small peltier heater/cooler that keeps the laser diode at a constant temperature. This ensures that the output wavelength remains as constant as possible (given variations in temperature and power).

Ambient light is the bane of structured light sensors. Again, there are measures put in place to mitigate this issue. One is an IR-pass filter at 830 nm over the IR camera. This prevents stray IR in other ranges (like from TV remotes and the like) from blinding the sensor or providing spurious results. However, even with this in place, the Kinect does not work well in places lit by sunlight. Sunlight's wide band IR has enough power in the 830 nm range to blind the sensor.

The distance at which the Kinect functions is also limited by the power of the laser diode. The laser diode power is limited by what is eye safe. Without the inclusion of the scattering filter, the laser diode in the Kinect is not eye safe; it's about 70 mW. This is why the scattering innovation by Prime Snese is so important: the extremely bright center dot is instead distributed amongst the 9 dots, allowing a higher powered laser diode to be used.

The IR camera operates at 30 Hz and pushes images out at 1200x960 pixels. These images are downsampled by the hardware, as the USB stack can't handle the transmission of that much data (combined with the RGB camera). The field of view in the system is 58 degrees horizontal, 45 degrees vertical, 70 degrees diagonal, and the operational range is between 0.8 meters and 3.5 meters. The resolution at 2 meters is 3 mm in X/Y and 1 cm in Z (depth). The camera itself is a MT9M001 by Micron, a monochrome camera with an active imaging array of 1280x1024 pixels, showing that the image is resized even before downsampling.

RGB Camera

The RGB camera, operating at 30 Hz, can push images at 640x512 pixels. The Kinect also has the option to switch the camera to high resolution, running at 15 frames per second (fps), which in reality is more like 10 fps at 1280x1024 pixels. Of course, the former is reduced slightly to match the depth camera; 640x480 pixels and 1280x1024 pixels are the outputs that are sent over USB. The camera itself possesses an excellent set of features including automatic white balancing, black reference, flicker avoidance, color saturation, and defect correction. The output of the RGB camera is bayered with a pattern of RG, GB. (Debayering is discussed in Chapter 3.)

Of course, neither the depth camera nor the RGB camera are of any use unless they're calibrated. While the XBox handles the calibration when the Kinect is connected to it, you need to take matters into your own hands. You'll be using an excellent piece of software by Nicholas Burrus called Kinect RGB Demo.

Kinect RGB Demo

You will use Kinect RGB Demo to calibrate your cameras. Why is this important?

- Without calibration, the cameras give us junk data. It's close, but close is often not good enough. If you want to see an example of this, post installation but precalibration, run RGBD-Reconstruction and see how poorly the scene matches between the color and the depth.

- Calibration is necessary to place objects in the world. Without calibration, the camera's intrinsic and extrinsic parameters (settings internal to and external to the camera, respectively) are factory set. When you calculate the position of a particular pixel in the camera's frame and translate it into the world frame, these parameters are how you do it. If they're not as close as you can get to correct, that pixel is misplaced in the world.

- Calibration will demonstrate some of the basic ideas behind what you'll be doing later. At its heart, calibration is about matching a known target to a set of possible images and calculating the differences. You will be matching unknown targets to possible images later, but many of the principles carry.

Let's get started.

Installation

The installation of RGB Demo isn't particularly difficult, but it can be problematic on certain platforms. We'll go step by step through how to install on each platform and we'll discuss workarounds for common issues. At the time of publication, the most recent version is found at http://labs.manctl.com/rgbdemo/.

Windows

First, you need to download RGB Demo from the link above; you'll want the source code. Unzip this into a source directory. We put it directly in C:/. You're going to be using MinGW and libfreenect.

1. Download and install QT for Windows from http://qt.nokia.com/downloads/windows-cpp and be sure to select to download http://get.qt.nokia.com/qt/source/qt-win-opensource-4.7.3-mingw.exe.

2. Add C:\Qt\4.7.3\bin (or wherever you installed to) to your path. Also, be sure to add two new system variables: QTDIR that lists where your core QT install is (typically C:\Qt\4.7.3), and QMAKESPEC set to win32-g++.

3. Run cmake-gui on your RGB Demo core folder, which is typically C:\RGBDemo-0.5.0-Source. Set the build directory to \build. See Figure 2-3 for a visual representation of the setup.

4. Generate the cmake file, then go to \build and run mingw32-make to build the system. The binaries will be in the \build\bin directory.

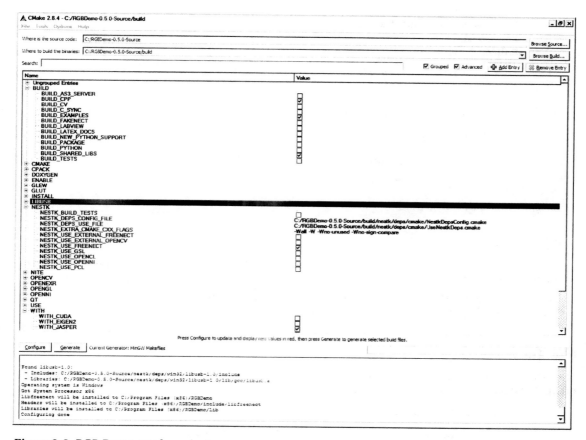

Figure 2-3. RGB Demo cmake-gui setup

Linux

The installations for both major flavors of Linux are almost exactly the same.

Ubuntu/Debian

1. Download the RGB Demo source and untar it into a directory of your choice. We used ~/kinect/.

2. Install the required packages for RGB Demo:

3. ```
sudo apt-get install libboost-all-dev libusb-1.0-0-dev libqt4-dev
libgtk2.0-dev cmake ccmake libglew1.5-dev libgs10-dev libglut3-dev
libxmu-dev
```
```

4. We recommend against using the prebuilt scripts to compile the source. Instead, run ccmake . on the RGBDemo-0.5.0-Source directory. Be sure to set the following flags:

- BUILD_EXAMPLES ON

- BUILD_FAKENECT ON

- BUILD_SHARED_LIBS ON

- NESTK_USE_FREENECT ON

- NESTK_USE_OPENNI OFF

- NESTK_USE_PCL OFF

5. Configure and generate, then run make.

Red Hat/Fedora

Red Hat and Fedora use the same installation procedure as Ubuntu, but run the prerequisite installation command as follows:

```
yum install libboost-all-dev libusb-1.0-0-dev libqt4-dev libgtk2.0-dev cmake ccmake
libglew1.5-dev libgs10-dev libglut3-dev libxmu-dev
```

Mac OS X

The installation for Mac OS X is relatively straightforward. The unified operating system makes the prebuilt installation scripts a snap to use.

1. Download and move RGB Demo into your desired directory.

2. Install QT from http://qt.nokia.com/downloads/qt-for-open-source-cpp-development-on-mac-os-x. We recommend using the Cocoa version if possible.

3. Once QT is installed, use your Terminal to move to the location of the RGB Demo, run tar xvfz RGBDemo-0.5.0-Source.tar.gz, cd into the directory, and run ./macosx_configuration.sh and ./macosx_build.sh scripts.

4. Once the system has finished installing, use Terminal to copy the calibrate_kinect_ir binary out of the .app folder in /build/bin/.

Making a Calibration Target

Now that your installation is complete, it' time to make a calibration target. From the base directory of your installation, go into /data/ and open chessboard_a4.pdf. You will use this chessboard to calibrate your cameras against a known target. Print it out on plain white paper. If you don't have A4 paper, change the size in your page setup until the entire target is clearly available. For standard letter paper, this was about an 80% reduction. After you've printed it out, tape it (using clear tape) to a piece of flat cardboard bigger than the target. Then, using a millimeter ruler, measure the square size, first to ensure that your squares truly are square and because you'll need that number for the calibration procedure (ours were 23 mm across). See Figure 2-4 for an example target.

Figure 2-4. Example calibration target

Calibrating with RGB Demo

Now that you have all the pieces of the system ready to go, let's calibrate!

1. Open `rgbd-viewer` in your `/build/bin` directory.

2. Turn on Dual IR/RGB Mode in the `Capture` directory.

3. Use Ctrl+G to capture images (we recommend at least 30). Figure 2-5 shows how the setup should look. Here are some calibration tips.

   - Get as close to the camera as possible with the target, but make sure the corners of the target are always in the image.

   - Split your images into two sets. For the first, make sure that the target is showing up at a distance from the camera (not blacked out); these are for the depth calibration. For the second, cover the IR emitter and get closer, as in the first point; these are for the stereo camera calibration.

- Get different angles and placements when you capture data. Pay special attention to the corners of the output window.

- Be sure your target is well lit but without specular reflections (bright spots).

4. Exit `rgbd-viewer` and execute `build/bin/calibrate_kinect_ir`. Be sure to set the following flags appropriately. An example output image is shown in Figure 2-6.

- `--pattern-size number_in_meters`

- `--input directory_where_your_images_are`

5. After the system runs, check the final pixel reprojection error. Less than 1 pixel is ideal.

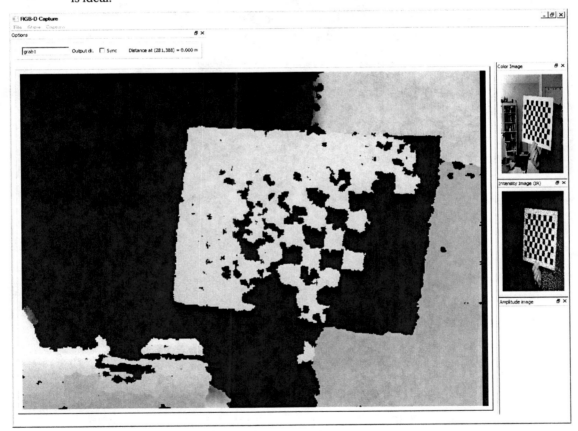

Figure 2-5. An example of a calibration position

Figure 2-6. An example of a corner detection in RGB Demo

Now that you have a kinect_calibration.yml file, you can run rgbd-viewer again, this time with the --calibration kinect_calibration.yml flag. Play around with it! You'll quickly notice a few things.

- Images are undistorted at the edges.

- Moving your mouse over a particular point in the image will give you its distance and it will be pretty accurate!

- You can apply some filters and see how they vary the output and system performance.

- By opening up 3D view in Show/3D View, you can see a 3D representation of the scene. Try out some of the filters when you're looking at this view.

So now what? You have a calibration file, kinect_calibration.yml, and software that uses it. This has nothing to do with your possible application, right? Untrue! The kinect_calibration.yml file is filled with useful information! Let's break it down. You can either track in your copy or look at Listing 2-1.

Listing 2-1. *kinect_calibration.yml*

```
%YAML:1.0
rgb_intrinsics: !!opencv-matrix
   rows: 3
   cols: 3
   dt: d
   data: [ 5.1849264445794347e+02, 0., 3.3438790034141709e+02, 0.,
       5.1589335524184094e+02, 2.5364152041171963e+02, 0., 0., 1. ]
rgb_distortion: !!opencv-matrix
   rows: 1
   cols: 5
   dt: d
   data: [ 2.4542340694293793e-01, -8.4327732173133640e-01,
       -1.8970692121976125e-03, 5.5458456701874270e-03,
       9.7254412755435449e-01 ]
depth_intrinsics: !!opencv-matrix
   rows: 3
   cols: 3
   dt: d
   data: [ 5.8089378818378600e+02, 0., 3.1345158291347678e+02, 0.,
       5.7926607093646408e+02, 2.4811989404941977e+02, 0., 0., 1. ]
depth_distortion: !!opencv-matrix
   rows: 1
   cols: 5
   dt: d
   data: [ -2.3987660910278472e-01, 1.5996260959757911e+00,
       -8.4261854767272721e-04, 1.1084546789468565e-03,
       -4.1018226565578777e+00 ]
R: !!opencv-matrix
   rows: 3
   cols: 3
   dt: d
   data: [ 9.9989106207829725e-01, -1.9337732418805845e-03,
       1.4632993438923941e-02, 1.9539514872675147e-03,
       9.9999715971478453e-01, -1.3647842134077237e-03,
       -1.4630312693856189e-02, 1.3932276959451122e-03,
       9.9989200060159855e-01 ]
T: !!opencv-matrix
   rows: 3
   cols: 1
   dt: d
   data: [ 1.9817238075432342e-02, -1.9169799354010252e-03,
       -2.7450591802116852e-03 ]
rgb_size: !!opencv-matrix
   rows: 1
   cols: 2
   dt: i
   data: [ 640, 480 ]
raw_rgb_size: !!opencv-matrix
   rows: 1
```

```
    cols: 2
    dt: i
    data: [ 640, 480 ]
depth_size: !!opencv-matrix
   rows: 1
   cols: 2
   dt: i
   data: [ 640, 480 ]
raw_depth_size: !!opencv-matrix
   rows: 1
   cols: 2
   dt: i
   data: [ 640, 480 ]
depth_base_and_offset: !!opencv-matrix
   rows: 1
   cols: 2
   dt: f
   data: [ 1.33541569e-01, 1.55009338e+03 ]
```

The values under rgb_intrinsics or depth_intrinsics are the camera's intrinsic parameters in pixel units. This is mapped in the matrices, as shown in Figure 2-7. Note that (c_x, c_y) is the principle point (usually the image center) while f_x and f_y are the focal lengths. What does all this mean? The cameras are calibrated using what is known as the pinhole model (reference Figure 2-8 for details). In short, the view of the scene is created by projecting a set of 3D points onto the image plane via a perspective transformation, and the following matrices are the way that projection happens.

Things aren't as simple as they look, though. Lenses also have distortion—mostly radial, but also a small amount of tangential. This is covered by the values k_1, k_2, k_3, p_1, p_2, where the k's are the radial distortion coefficients, and the p's are the tangential. These are stored in rgb_distortion and depth_distortion; see Figure 2-9.

$$\begin{bmatrix} f_x & 0 & c_x \\ 0 & f_y & c_y \\ 0 & 0 & 1 \end{bmatrix}$$

$c_x = 334.388...$ 　　　　　　　$f_x = 518.492...$

$c_y = 253.641...$ 　　　　　　　$f_y = 515.893...$

Figure 2-7. The camera matrix and example values for rgb_intrinsics

21

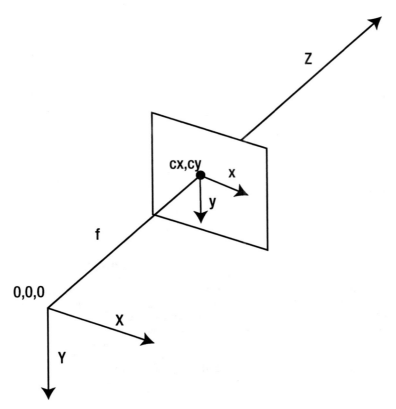

Figure 2-8. *The pinhole model (from the ros.org wiki)*

$$\begin{bmatrix} k_1 & k_2 & P_1 & P_2 & k_3 \end{bmatrix}$$

$k_1 = 0.24542...$ $P_1 = -0.00189...$

$k_2 = -0.84327...$ $P_2 = -0.00554...$

$k_3 = 0.97254...$

Figure 2-9. *The distortion matrix and example values for rgb_distortion*

Finally, there is one more important set of values, R and T. In truth, these are separated only to make it easier to read. They work together to translate and rotate the projected point from the world into a coordinate frame in reference to the camera. The traditional format is shown in Figure 2-10.

$$\begin{bmatrix} r_11 & r_12 & r_13 & t_1 \\ r_21 & r_22 & r_23 & t_2 \\ r_31 & r_32 & r_33 & t_3 \end{bmatrix}$$

Figure 2-10. *Combined R|T matrix*

That's quite a bit of information to take in, but it's only a quick overview. Searching on the Web will yield way more information than can be put into this book. In particular, check out http://opencv.willowgarage.com/documentation/cpp/camera_calibration_and_3d_reconstruction.html for an in-depth explanation.

Tilting Head and Accelerometer

The Kinect hides two inter-related and important systems inside: a method to tilt the head of the Kinect up and down plus an accelerometer. The head tilting is relatively simple; it's a motor with some gearing to drive the head up and down. Take a look at Figure 2-11 to see how it is constructed. One thing that the system does not have is a way to determine what position the head is in; that requires the accelerometer.

An accelerometer, at its most simple, is a device that measures acceleration. In the case of a fixed system like the Kinect, the accelerometer tells the system which way is down by measuring the acceleration due to gravity. This allows the system to set its head at exactly level and to calibrate to a value so the head can be moved at specific angles.

However, the accelerometer can be used for much more. Nicholas Burress built an application that takes advantage of this situation to actively build a scene when you move the Kinect around through an area. Play around with rgdb-reconstruction and you can develop a scene like you see in Figure 2-12.

As for the accelerometer itself, it is a KXSD9-2050. Factory set at +- 2 g's, it has a sensitivity of 819 counts per g.

Figure 2-11. Kinect base teardown from www.ifixit.com/Teardown/Microsoft-Kinect-Teardown/4066/1

Figure 2-12. 3D scene reconstruction via rgbd-reconstructor

VOLUMETRIC SENSING

Time for your first project, a volumetric sensor hooked up via hardware to a light. Simple, eh? Actually, unlike motion detectors, your system can't be fooled by moving slowly, and unlike IR sensors, it can't be fooled by heating up the room (see Sneakers). So you'll have a highly effective, if somewhat clunky, alarm system in the end.

We're going to dump a huge amount of code on you. Most of it will be pretty incomprehensible right now, so you'll have to trust that it will be all explained later.

Required Parts

- Arduino Uno

- A wiring breakout board

- A lamp to sacrifice to the project

- A relay that triggers on 5 V that can handle enough wattage to run the light bulb (we used an Omron G6a-274P-ST-US, which is a 5V relay rated to 0.6 A at 125 V)

First things first, let's get some of the necessary library code out of the way.

Create a new directory called OKFlower.

Run git clone git://gitorious.org/serial-port/serial-port.git inside this directory.

Copy AsyncSerial.cpp and .h as well as BufferedAsyncSerial.cpp and .h into your OKFlower directory.

Create the file in Listing 2-2, OKFlower.cpp, in the OKFlower directory.

Listing 2-2. OKFlower.cpp

```
/*
 * This file is part of the OpenKinect Project. http://www.openkinect.org
 *
 * Copyright (c) 2010 individual OpenKinect contributors. See the CONTRIB file
 * for details.
 *
 * This code is licensed to you under the terms of the Apache License, version
 * 2.0, or, at your option, the terms of the GNU General Public License,
 * version 2.0. See the APACHE20 and GPL2 files for the text of the licenses,
 * or the following URLs:
 * http://www.apache.org/licenses/LICENSE-2.0
```

```
* http://www.gnu.org/licenses/gpl-2.0.txt
*
* If you redistribute this file in source form, modified or unmodified, you
* may:
*    1) Leave this header intact and distribute it under the same terms,
*       accompanying it with the APACHE20 and GPL20 files, or
*    2) Delete the Apache 2.0 clause and accompany it with the GPL2 file, or
*    3) Delete the GPL v2 clause and accompany it with the APACHE20 file
* In all cases you must keep the copyright notice intact and include a copy
* of the CONTRIB file.
*
* Binary distributions must follow the binary distribution requirements of
* either License.
*/

#include <iostream>
#include <libfreenect.hpp>
#include <pthread.h>
#include <stdio.h>
#include <string.h>
#include <cmath>
#include <vector>
#include <ctime>
#include <boost/thread/thread.hpp>
#include "pcl/common/common_headers.h"
#include "pcl/features/normal_3d.h"
#include "pcl/io/pcd_io.h"
#include "pcl/visualization/pcl_visualizer.h"
#include "pcl/console/parse.h"
#include "pcl/point_types.h"
#include <pcl/kdtree/kdtree_flann.h>
#include <pcl/surface/mls.h>
#include "boost/lexical_cast.hpp"
#include "pcl/filters/voxel_grid.h"
#include "pcl/octree/octree.h"
#include "BufferedAsyncSerial.h"

using namespace std;
using namespace boost;

///Mutex Class
class Mutex {
public:
      Mutex() {
             pthread_mutex_init( &m_mutex, NULL );
      }
```

```
void lock() {
        pthread_mutex_lock( &m_mutex );
}
void unlock() {
        pthread_mutex_unlock( &m_mutex );
}

class ScopedLock
{
        Mutex & _mutex;
public:
        ScopedLock(Mutex & mutex)
                : _mutex(mutex)
        {
                _mutex.lock();
        }
        ~ScopedLock()
        {
                _mutex.unlock();
        }
};
private:
        pthread_mutex_t m_mutex;
};


///Kinect Hardware Connection Class
/* thanks to Yoda---- from IRC */
class MyFreenectDevice : public Freenect::FreenectDevice {
public:
        MyFreenectDevice(freenect_context *_ctx, int _index)
                : Freenect::FreenectDevice(_ctx, _index),
depth(freenect_find_depth_mode(FREENECT_RESOLUTION_MEDIUM,
FREENECT_DEPTH_REGISTERED).bytes),m_buffer_video(freenect_find_video_mode(FREENEC
T_RESOLUTION_MEDIUM, FREENECT_VIDEO_RGB).bytes), m_new_rgb_frame(false),
m_new_depth_frame(false)
        {

        }
        //~MyFreenectDevice(){}
        // Do not call directly even in child
        void VideoCallback(void* _rgb, uint32_t timestamp) {
                Mutex::ScopedLock lock(m_rgb_mutex);
                uint8_t* rgb = static_cast<uint8_t*>(_rgb);
                std::copy(rgb, rgb+getVideoBufferSize(), m_buffer_video.begin());
                m_new_rgb_frame = true;
```

```cpp
    };
    // Do not call directly even in child
    void DepthCallback(void* _depth, uint32_t timestamp) {
            Mutex::ScopedLock lock(m_depth_mutex);
            depth.clear();
            uint16_t* call_depth = static_cast<uint16_t*>(_depth);
            for (size_t i = 0; i < 640*480 ; i++) {
                    depth.push_back(call_depth[i]);
            }
            m_new_depth_frame = true;
    }
    bool getRGB(std::vector<uint8_t> &buffer) {
            //printf("Getting RGB!\n");
            Mutex::ScopedLock lock(m_rgb_mutex);
            if (!m_new_rgb_frame) {
                    //printf("No new RGB Frame.\n");
                    return false;
            }
            buffer.swap(m_buffer_video);
            m_new_rgb_frame = false;
            return true;
    }

    bool getDepth(std::vector<uint16_t> &buffer) {
            Mutex::ScopedLock lock(m_depth_mutex);
            if (!m_new_depth_frame)
                    return false;
            buffer.swap(depth);
            m_new_depth_frame = false;
            return true;
    }

private:
    std::vector<uint16_t> depth;
    std::vector<uint8_t> m_buffer_video;
    Mutex m_rgb_mutex;
    Mutex m_depth_mutex;
    bool m_new_rgb_frame;
    bool m_new_depth_frame;
};


///Start the PCL/OK Bridging

//OK
Freenect::Freenect freenect;
```

```
MyFreenectDevice* device;
freenect_video_format requested_format(FREENECT_VIDEO_RGB);
double freenect_angle(0);
int got_frames(0),window(0);
int g_argc;
char **g_argv;
int user_data = 0;


//PCL
pcl::PointCloud<pcl::PointXYZRGB>::Ptr cloud (new
pcl::PointCloud<pcl::PointXYZRGB>);
pcl::PointCloud<pcl::PointXYZRGB>::Ptr bgcloud (new
pcl::PointCloud<pcl::PointXYZRGB>);
pcl::PointCloud<pcl::PointXYZRGB>::Ptr voxcloud (new
pcl::PointCloud<pcl::PointXYZRGB>);
float resolution = 50.0;
// Instantiate octree-based point cloud change detection class
pcl::octree::OctreePointCloudChangeDetector<pcl::PointXYZRGB> octree
(resolution);

bool BackgroundSub = false;
bool hasBackground = false;
bool Voxelize = false;
unsigned int voxelsize = 1; //in mm
unsigned int cloud_id = 0;



///Keyboard Event Tracking
void keyboardEventOccurred (const pcl::visualization::KeyboardEvent &event,
                           void* viewer_void)
{
  boost::shared_ptr<pcl::visualization::PCLVisualizer> viewer =
*static_cast<boost::shared_ptr<pcl::visualization::PCLVisualizer> *>
(viewer_void);
  if (event.getKeySym () == "c" && event.keyDown ())
  {
    std::cout << "c was pressed => capturing a pointcloud" << std::endl;
    std::string filename = "KinectCap";
    filename.append(boost::lexical_cast<std::string>(cloud_id));
    filename.append(".pcd");
    pcl::io::savePCDFileASCII (filename, *cloud);
    cloud_id++;
  }
```

```
if (event.getKeySym () == "b" && event.keyDown ())
{
    std::cout << "b was pressed" << std::endl;
    if (BackgroundSub == false)
    {
        //Start background subtraction
        if (hasBackground == false)
        {
            //Copy over the current cloud as a BG cloud.
            pcl::copyPointCloud(*cloud, *bgcloud);
            hasBackground = true;
        }
        BackgroundSub = true;
    }
    else
    {
        //Stop Background Subtraction
        BackgroundSub = false;
    }
}

if (event.getKeySym () == "v" && event.keyDown ())
{
    std::cout << "v was pressed" << std::endl;
    Voxelize = !Voxelize;
}

}


// --------------
// -----Main-----
// --------------
int main (int argc, char** argv)
{
    if (argc != 4)
    {
        cerr << "Usage: OKFlower <baud> <device> <percentage>\n";
        cerr << "Typ: 9600 /dev/tty.usbmodemfd131 10\n";
        return 1;
    }

    //Percentage Change
    float percentage = boost::lexical_cast<float>(argv[3]);
    percentage = percentage/100.0;
```

```
//More Kinect Setup
static std::vector<uint16_t> mdepth(640*480);
static std::vector<uint8_t> mrgb(640*480*4);

// Fill in the cloud data
cloud->width    = 640;
cloud->height   = 480;
cloud->is_dense = false;
cloud->points.resize (cloud->width * cloud->height);

// Create and setup the viewer
boost::shared_ptr<pcl::visualization::PCLVisualizer> viewer (new
pcl::visualization::PCLVisualizer ("3D Viewer"));
    viewer->registerKeyboardCallback (keyboardEventOccurred, (void*)&viewer);
    viewer->setBackgroundColor (255, 255, 255);
    viewer->addPointCloud<pcl::PointXYZRGB> (cloud, "Kinect Cloud");
    viewer->setPointCloudRenderingProperties
(pcl::visualization::PCL_VISUALIZER_POINT_SIZE, 1, "Kinect Cloud");
    viewer->addCoordinateSystem (1.0);
    viewer->initCameraParameters ();

//Voxelizer Setup
pcl::VoxelGrid<pcl::PointXYZRGB> vox;


device = &freenect.createDevice<MyFreenectDevice>(0);
device->startVideo();
device->startDepth();
boost::this_thread::sleep (boost::posix_time::seconds (1));
//Grab until clean returns
int DepthCount = 0;
while (DepthCount == 0) {
        device->updateState();
        device->getDepth(mdepth);
        device->getRGB(mrgb);
        for (size_t i = 0;i < 480*640;i++)
                DepthCount+=mdepth[i];
}

device->setVideoFormat(requested_format);
//-------------------
// -----Main loop-----
//-------------------
double x = NULL;
double y = NULL;
int iRealDepth = 0;
```

31

```
        try {
                //BufferedAsync Setup
                printf("Starting up the BufferedAsync Stuff\n");
        BufferedAsyncSerial serial(argv[2], boost::lexical_cast<unsigned
int>(argv[1]));

        while (!viewer->wasStopped ())
        {
                device->updateState();
                device->getDepth(mdepth);
                device->getRGB(mrgb);

            size_t i = 0;
        size_t cinput = 0;
        for (size_t v=0 ; v<480 ; v++)
        {
                for ( size_t u=0 ; u<640 ; u++, i++)
                {
                        //pcl::PointXYZRGB result;
                        iRealDepth = mdepth[i];
                                //DepthCount+=iRealDepth;
                                //printf("fRealDepth = %f\n",fRealDepth);
                                //fflush(stdout);
                                freenect_camera_to_world(device->getDevice(), u, v,
iRealDepth, &x, &y);
                                cloud->points[i].x  = x;//1000.0;
                                cloud->points[i].y  = y;//1000.0;
                    cloud->points[i].z = iRealDepth;//1000.0;
                cloud->points[i].r = mrgb[i*3];
                    cloud->points[i].g = mrgb[(i*3)+1];
                 cloud->points[i].b = mrgb[(i*3)+2];
                    //cloud->points[i] = result;
                //printf("x,y,z = %f,%f,%f\n",x,y,iRealDepth);
                //printf("RGB = %d,%d,%d\n",
mrgb[i*3],mrgb[(i*3)+1],mrgb[(i*3)+2]);
                }
                }

                if (BackgroundSub) {
                        pcl::PointCloud<pcl::PointXYZRGB>::Ptr fgcloud (new
pcl::PointCloud<pcl::PointXYZRGB>);
                        octree.deleteCurrentBuffer();

                        // Add points from background to octree
                        octree.setInputCloud (bgcloud);
                        octree.addPointsFromInputCloud ();
```

```
                    // Switch octree buffers: This resets octree but keeps
previous tree structure in memory.
                    octree.switchBuffers ();

                    // Add points from the mixed data to octree
                    octree.setInputCloud (cloud);
                    octree.addPointsFromInputCloud ();

                    std::vector<int> newPointIdxVector;

                    // Get vector of point indices from octree voxels which did
not exist in previous buffer
                    octree.getPointIndicesFromNewVoxels (newPointIdxVector);

                    for (size_t i = 0; i < newPointIdxVector.size(); ++i) {
                            fgcloud->push_back(cloud-
>points[newPointIdxVector[i]]);
                    }

                    viewer->updatePointCloud (fgcloud, "Kinect Cloud");

                    printf("Cloud Size: %d vs. Max: %d vs. perc: %f\n",
newPointIdxVector.size(), 640*480,
(float)newPointIdxVector.size()/(float)(640.0*480.0));
                    if (((float)newPointIdxVector.size()/(float)(640.0*480.0))
>= percentage) {
                            printf("Person in Scene!\n");
                            serial.writeString("L");
                    }
                    else
                            serial.writeString("O");
            }
            else if (Voxelize) {
                    vox.setInputCloud (cloud);
                    vox.setLeafSize (5.0f, 5.0f, 5.0f);
                    vox.filter (*voxcloud);
                    viewer->updatePointCloud (voxcloud, "Kinect Cloud");
            }
            else
            viewer->updatePointCloud (cloud, "Kinect Cloud");

            viewer->spinOnce ();
    }
    serial.close();
    } catch(boost::system::system_error& e)
```

```
    {
            cout<<"Error: "<<e.what()<<endl;
            device->stopVideo();
            device->stopDepth();
            return 1;
    }
    device->stopVideo();
    device->stopDepth();
    return 0;
}
```

This code is an extension of the code we discussed in the previous chapter. It adds a feature that we'll discuss in Chapter 9, background subtraction. Create a CMakeLists.txt file with the following inside of it (Listing 2-3), and then compile by running ccmake . Following the basic configure, generate and exit.

Listing 2-3. CMakeLists.txt

```
cmake_minimum_required(VERSION 2.8 FATAL_ERROR)
set(CMAKE_C_FLAGS "-Wall")

project(OKFlower)

find_package(PCL 1.4 REQUIRED)

include_directories(${PCL_INCLUDE_DIRS})
link_directories(${PCL_LIBRARY_DIRS})
add_definitions(${PCL_DEFINITIONS})


if (WIN32)
  set(THREADS_USE_PTHREADS_WIN32 true)
  find_package(Threads REQUIRED)

  include_directories(${THREADS_PTHREADS_INCLUDE_DIR})
endif()

include_directories(. ${PCL_INCLUDE_DIRS})
link_directories(${PCL_LIBRARY_DIRS})
add_definitions(${PCL_DEFINITIONS})

add_executable(OKFlower OKFlower.cpp AsyncSerial.cpp BufferedAsyncSerial.cpp)

if(APPLE)
  target_link_libraries(OKFlower freenect ${PCL_LIBRARIES})
else()
  find_package(Threads REQUIRED)
  include_directories(${USB_INCLUDE_DIRS})

  if (WIN32)
```

```
    set(MATH_LIB "")
  else(WIN32)
    set(MATH_LIB "m")
  endif()
  target_link_libraries(OKFlower freenect ${CMAKE_THREAD_LIBS_INIT} ${MATH_LIB}
${PCL_LIBRARIES})
endif()
```

Great! You've now compiled the code that will interface with the Arduino. The Arduino itself is easy to use: simply download the Arduino IDE from arduino.cc, hook your Arduino up to your machine, and see it run. You're going to upload the sketch (program) in Listing 2-4 into the Arduino to accept the serial communications that you're sending from the OKFlower program.

Listing 2-4. Arduino Sketch

```
int inByte = 0;              // incoming serial byte

void setup()
{
  // start serial port at 9600 bps:
  Serial.begin(9600);
  pinMode(12, OUTPUT);    // relay is on output
  pinMode(13, OUTPUT); //Set LED to monitor on Uno
}

void loop()
{
  // if we get a valid byte, read analog ins:
  if (Serial.available() > 0) {
    // get incoming byte:
    inByte = Serial.read();
    char letter = char(inByte);
    if (letter == 'L') {
      digitalWrite(12, HIGH);
      digitalWrite(13, HIGH);
    }
    else
    {
      digitalWrite(12, LOW);
      digitalWrite(13, LOW);
    }
  }
}
```

Great! That code is very simple. It will make both pin 12 and pin 13 high (on) when the serial system receives an L. Otherwise, it will make them low (off). You can test this in the serial terminal inside of the IDE, as shown in Figure 2-13. Be sure to note the port name and baud rate when you upload your code; you'll need it to start OKFlower.

The next step is to physically wire up the system. Thankfully, this is exceedingly simple. Every relay has a NO (normally open) connection—the place where you attach your hot leads from the socket (on one side) and the light (on the other). One example of this relay wiring is shown in Figure 2-14. Your grounds are tied together, as shown in Figure 2-14. The Arduino is attached to the positive trigger pin with a wire to pin 12 on the Arduino and a negative pin to the Arduino ground, as shown in Figure 2-15.

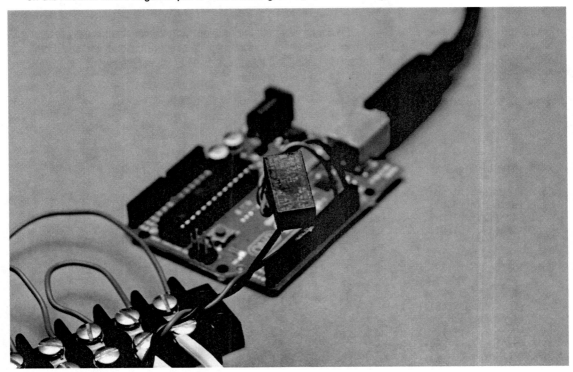

Figure 2-13. Relay wiring example

Figure 2-14. Block wiring example. The white wires go to the lamp and the socket. The long green wire ties their grounds together. The short green wires force the flow of electricity to go through the relay.

Figure 2-15. Wiring into the Arduino itself. The red wire is hot and is in pin 12. The black is ground and is in pin 14.

Great! Now that everything is wired up, you can try it. Start OKFlower with the following arguments: `OKFlower <PORT> <BAUDRATE> <PERCENTAGE>` and be sure to type them in correctly! Percentage is a tunable parameter based on how close your Kinect is to your area of interest. The higher the percentage, the more of the scene that needs to be filled to trigger the alarm.

If everything works, the Kinect point cloud should come up and you can drag it around to look at it. When you're ready (and out of the scene!) hit b to subtract the background. Now, when anything enters back into the scene that is different enough from the captured background, it will trigger the light! See Figure 2-16 for the final design.

Figure 2-16. A lit alarm light. Obviously someone (or something) has been detected!

Great! Nicely done. Here are a few possible expansions to consider.

- Add some hysteresis to prevent the light from flickering when passing close to the edge of the percentage.

- Add some motors or more complicated systems to the Arduino to trigger when people are seen.

- Add another output to the computer side. Perhaps it should Tweet you when someone is detected?

Good luck!

Summary

Chapter 2 is done! You've learned how the Kinect's depth sensing works, some more information about cameras in general, and how to calibrate your Kinect. You've also built a working volumetric alarm system, so you now understand far more about the hardware of the Kinect than most people.

CHAPTER 3

Software

In this chapter, we willdiscuss the current options for Kinect drivers and explain why OpenKinect is the best choice. We'll also walk through installing OpenCV (Open Computer Vision) and the PCL (Point Cloud Library)—the two mathematics packages we will be using to work with Kinect data. Finally, we'll implement the OpenKinect-PCL bridge that will a basis for future work.

Let's get started!

Exploring the Kinect Drivers

If you've been following the news on the Kinect or seen some of the different systems available, you've probably heard about a few of the drivers we're going to discuss. While most would do a fine job at teaching you how to use a depth sensor in your code, we're more interested in the *best* option.

OpenNI

OpenNI is the "industry-led, not-for-profit organization" that publishes the middleware for the Kinect, as well as a basic driver. Of the organizations involved, the two key players are Willow Garage and PrimeSense. OpenNI works seamlessly with PCL and even has a workable middleware that handles things like skeletonization. So why not use OpenNI? Sadly, the code is not open source, and the lack of support is worrying.

Microsoft Kinect SDK

Released on June 16, 2011, the Microsoft Kinect SDK only functions on Windows 7 and is only available for noncommercial uses. The SDK is not open source, and although it possesses many features, it lacks some basics for such a heavyweight SDK (things like gesture recognition and hand detection). The newest version of the Microsoft Kinect SDK was just released. It specifies that you can only use it for commercial uses when you're using the Kinect for Windows (K4W), which we don't cover here.

OpenKinect

Héctor Martin is widely credited with the initial hack required to create the first open source driver to create applications for the Kinect. In truth, he was part of a larger movementdriven by the desire to experiment, Adafruit Industries' bounty (which reached $3,000), and Microsoft's stated threat that it would "work closely with law enforcement . . . to keep the Kinect tamper resistant." The Kinect was released on November 4. Four days later, a hacker by the name of AlexP had reversed engineered it but refused to release his code unless the community put together $10,000 to fund his research. While other hackers were still working on getting the Kinect working on the November 9, Héctor blew past everyone when the Kinect was released in Spain on November 10th. At 10 a.m., he bought his Kinect.At 11 a.m., he caught up with the current level of work. At noon, he claimed victory, demonstrating video and depth data streaming from the device.

Since then, the OpenKinect project has evolved based on Héctor's initial hack– it supports motor, LED, and camera control, as well as accelerometer output. There are many wrappers for your favorite programming languages. While OpenKinect is currently a low-level driver, its strong development community is working toward bringing out many of the features people are asking for. You'll be developing some of these features in this book. The community is extremely supportive; the system is totally open source, and you can use it in any project.

For these reasons, we'll be using OpenKinect as the basis for all of our projects.

Installing OpenCV

OpenCV is a prebuilt library of real-time computer vision functions. It has over 2,000 functions dedicated to a wide variety of computer vision problems—fitting, camera calibration (it is the basis of the RGB-Demo calibration we did in the preceding chapter), segmentation, tracking, transforms, and more. OpenCVwill be an immense help to implement our system, and if we use a function from OpenCV, you can be sure that we will explain the function clearly and concisely, so you can return to the toolkit again and again.

OpenCV is relatively straightforward to install, but it does have quite a few dependencies. Lucky for us that we already installed most of them! The next few sections go over how to install OpenCV on Windows, Linux, and the Mac.

Windows

Windows is again the hardest platform to install for. We'll be working through a MinGW installation, in keeping with our focus on open source systems and tools. To install OpenCV on Windows, we'll need to build it from source. Download the source code, put it into a directory structure. Compile. The full-blown process is as follows:

1. Go to http://threadingbuildingblocks.org and get the most recent version of Threading Building Blocks (TBB). This will allow us to unlock the parallel processing algorithms for OpenCV. Be sure to select the one ending in _win.zip; the others won't work.

2. Unzip the TBB download into a directory of your choice.

3. Head over to http://opencv.willowgarage.com, and click the Windows option in the Downloads section. On the next page, be sure to get the version ending in –win.zip;the other one is set up for Visual Studio 2010, and we're using MinGW.

4. Unzip the OpenCV download into a directory of your choice.

5. Create a new folder in C:\ called OpenCV2.2MinGW; this will be the directory into which we build.

6. Run cmake-gui.Set both the source (the place you unzipped OpenCV) and the build (the directory created in step 5) correctly.Configure.

7. Select TBB to enable it, andconfigure again.

8. Set the TBB paths correctly, andconfigure again.

9. At this point, your installationshould look like Figure 3-1.

10. Generate and close cmake-gui.

11. Open a command prompt, head over to your build directory (should be C:\OpenCV2.2MinGW), and type the command mingw32-make.

12. After the make command finishes, type the mingw32-make install command.

13. Add C:\OpenCV2.2MinGW\bin to the PATHin your environmental variables, and close your command prompt.

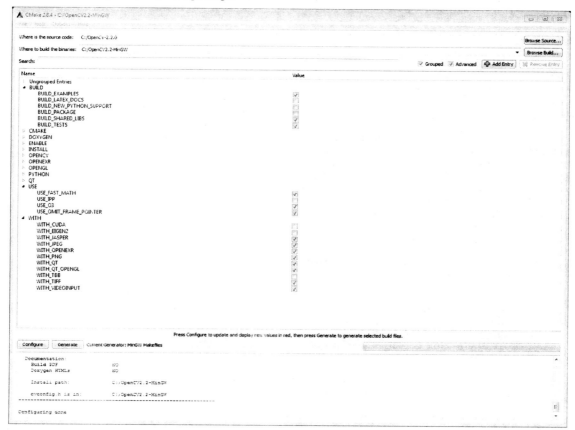

Figure 3-1.*The OpenCV cmake-gui configuration screen*

Linux

Before installing OpenCV to a Linux machine, you'll need to install the following prerequisites. The method variesamong Linux distributions, but the libraries are the same.

- GTK+ 2.0 or better.Be sure to include all headers and development files

- `libpng`, `zlib`, `libjpeg`, `libtiff`, and `libjasper` with all development files

- SWIG 1.3.30 or better

- `ffmpeg` and `libavcodec` from `0.4.9-pre1` or later, plus the headers and development files

- `libdc1394` 2.0 or better, plus theheaders

See Listing 3-1 for the Ubuntu installation commands. Execute those commands on your system to install the preceding list of prerequisites. If you are running a different distribution of Linux, follow whatever install process is correct for what you have.

Listing 3-1. Ubuntu OpenCV Installation Commands

```
sudo apt-get install libpng12-0 libpng12-dev libpng++-dev libpng3
sudo apt-get install libpnglite-dev libpngwriter0-dev libpngwriter0c2
sudo apt-get install zlib1g-dbg zlib1g zlib1g-dev
sudo apt-get install libjasper-dev libjasper-runtime libjasper1
sudo apt-get install pngtools libtiff4-dev libtiff4 libtiffxx0c2 libtiff-tools
sudo apt-get install libjpeg62 libjpeg62-dev libjpeg62-dbg libjpeg-progs
sudo apt-get install ffmpeg libavcodec-dev libavcodec52 libavformat52 libavformat-dev
sudo apt-get install libgstreamer0.10-0-dbg libgstreamer0.10-0  libgstreamer0.10-dev
sudo apt-get install libxine1-ffmpeg  libxine-dev libxine1-bin
sudo apt-get install libunicap2 libunicap2-dev
sudo apt-get install libdc1394-22-dev libdc1394-22 libdc1394-utils
sudo apt-get install swig
sudo apt-get install libv4l-0 libv4l-dev
sudo apt-get install python-numpy
sudo apt-get install libtbb2 libtbb-dev
```

With the prerequisites out of the way, you can download and install the OpenCV software itself. Here are the steps to follow:

1. Download the stable binaries from `http://opencv.willowgarage.com`, and click on the Unix option in the Downloads section. Download and uncompress the tar file for the most recent stable version.

2. Create a `build` directory in the `OpenCV` folder. Then, use thecdcommand to go to the newly created directory, and run ccmake ...

3. Configure and set up the options as shown in Figure 3-2. Be sure to turn off CUDA, EIGEN,and IPP and to turn on TBB.

4. Generate and exit ccmake, and run `make`.

5. After the build finishes, run `sudo make install` to put OpenCV in the right place and add it to your library path.

6. In the `OpenCV` folder, use cdto go into the build directory and again to go into bin.Run the OpenCV tests with `./opencv_test_core`.The results should come back stating that OpenCV is fully operational.

```
                                    Page 1 of 2
BUILD_DOCS                      ON
BUILD_EXAMPLES                  ON
BUILD_NEW_PYTHON_SUPPORT        ON
BUILD_PACKAGE                   OFF
BUILD_SHARED_LIBS               ON
BUILD_TESTS                     ON
BZIP2_LIBRARIES                 /usr/lib/libbz2.so
CMAKE_BACKWARDS_COMPATIBILITY   2.4
CMAKE_BUILD_TYPE
CMAKE_CONFIGURATION_TYPES       Debug;Release
CMAKE_INSTALL_PREFIX            /usr/local
CMAKE_VERBOSE                   OFF
CUDA_BUILD_CUBIN                OFF
CUDA_BUILD_EMULATION            OFF
CUDA_SDK_ROOT_DIR               CUDA_SDK_ROOT_DIR-NOTFOUND
CUDA_TOOLKIT_ROOT_DIR           CUDA_TOOLKIT_ROOT_DIR-NOTFOUND
CUDA_VERBOSE_BUILD              OFF
EIGEN_INCLUDE_PATH              EIGEN_INCLUDE_PATH-NOTFOUND
ENABLE_PROFILING                OFF
ENABLE_SSE                      ON
ENABLE_SSE2                     ON
ENABLE_SSE3                     OFF
ENABLE_SSE41                    OFF
ENABLE_SSE42                    OFF
ENABLE_SSSE3                    OFF
EXECUTABLE_OUTPUT_PATH          /home/jkramer/kinect/OpenCV-2.3.0/build/bin
INSTALL_C_EXAMPLES              OFF
INSTALL_PYTHON_EXAMPLES         OFF
LIBRARY_OUTPUT_PATH             /home/jkramer/kinect/OpenCV-2.3.0/build/lib
OPENCV_BUILD_3RDPARTY_LIBS      OFF
OPENCV_CONFIG_FILE_INCLUDE_DIR  /home/jkramer/kinect/OpenCV-2.3.0/build
OPENCV_EXTRA_C_FLAGS
OPENCV_EXTRA_C_FLAGS_DEBUG
OPENCV_EXTRA_C_FLAGS_RELEASE
OPENCV_EXTRA_EXE_LINKER_FLAGS
OPENCV_EXTRA_EXE_LINKER_FLAGS_
OPENCV_EXTRA_EXE_LINKER_FLAGS_
OPENCV_WARNINGS_ARE_ERRORS      OFF
OPENEXR_INCLUDE_PATH            OPENEXR_INCLUDE_PATH-NOTFOUND
PVAPI_INCLUDE_PATH              PVAPI_INCLUDE_PATH-NOTFOUND
PYTHON_PACKAGES_PATH            lib/python2.6/dist-packages
USE_FAST_MATH                   ON
USE_OMIT_FRAME_POINTER          ON
USE_PRECOMPILED_HEADERS         ON
WITH_1394                       ON
WITH_CUDA                       OFF
WITH_EIGEN                      OFF
WITH_FFMPEG                     ON
WITH_GSTREAMER                  ON
WITH_GTK                        ON

BUILD DOCS: Build OpenCV Reference Manual
Press [enter] to edit option
Press [c] to configure
Press [h] for help        Press [q] to quit without generating
Press [t] to toggle advanced mode (Currently Off)
```

Figure 3-2.Cmake in Ubuntu for OpenCV

45

Mac OS X

While there is a MacPorts version of OpenCV available (2.2.0), we're after the most recent stable code (at the time of this writing, 2.3rc). Most of the installation for OS X mirrors that for Linux, with some caveats. Here is the process to follow:

1. Go to http://threadingbuildingblocks.org, and get the most recent version of Threading Building Blocks (TBB). This will allow us to unlock the parallel processing algorithms for OpenCV. Under Downloads Stable Release, pick the newest option and be sure to download the one that ends in _mac.tgz, because the others won't work.

2. Uncompress the tar file into the directory of your choice.

3. In your home directory (cd~), edit your .profile file (using open .profile). In the TBB directory created in step 2, use the cdcommand to enter the bin directory, and open tbbvars.sh. Copy all of the text below the introductory line into the bottom of your .profile file, making sure to edit the install directory to match the full path to the TBB directory. Save and close the file.Close and reopen your terminal to set it up.

4. Download the stable binaries from http://opencv.willowgarage.com and click on the Unix option in the Downloads section. Download and uncompress the tar file for the most recent stable version.

5. Create a builddirectory in the OpenCV folder. Then, use cdto go to the newly created directory and run ccmake ...

6. Configure and set up the options as shown in Figure 3-3. Be sure to turn off CUDA, EIGEN and IPP, and to turn on TBB.

7. Generate and exit ccmake, and then run make.

8. After the build finishes, run sudo make install to put OpenCV in the right place and add it to your library path.

9. In the OpenCV folder, use cdto go into the build directory and then into bin. Run the OpenCV tests with ./opencv_test_core. It should come back fully operational.

```
BUILD_DOXYGEN_DOCS                  OFF
BUILD_EXAMPLES                      OFF
BUILD_LATEX_DOCS                    OFF
BUILD_NEW_PYTHON_SUPPORT            ON
BUILD_PACKAGE                       OFF
BUILD_SHARED_LIBS                   ON
BUILD_TESTS                         ON
BZIP2_LIBRARIES                     /usr/lib/libbz2.dylib
CMAKE_BACKWARDS_COMPATIBILITY       2.4
CMAKE_BUILD_TYPE
CMAKE_CONFIGURATION_TYPES           Debug;Release
CMAKE_INSTALL_PREFIX                /usr/local
CMAKE_OSX_ARCHITECTURES
CMAKE_OSX_DEPLOYMENT_TARGET
CMAKE_OSX_SYSROOT                   /Developer/SDKs/MacOSX10.6.sdk
CMAKE_VERBOSE                       OFF
EIGEN2_INCLUDE_PATH                 EIGEN2_INCLUDE_PATH-NOTFOUND
ENABLE_PROFILING                    OFF
ENABLE_SSE                          ON
ENABLE_SSE2                         ON
ENABLE_SSE3                         OFF
ENABLE_SSE41                        OFF
ENABLE_SSE42                        OFF
ENABLE_SSSE3                        OFF
EXECUTABLE_OUTPUT_PATH              /Users/Sysop/OpenCV-2.2.0/build/bin
INSTALL_C_EXAMPLES                  OFF
INSTALL_PYTHON_EXAMPLES             OFF
IPP_PATH                            IPP_PATH-NOTFOUND
LIBRARY_OUTPUT_PATH                 /Users/Sysop/OpenCV-2.2.0/build/lib
OPENCV_BUILD_3RDPARTY_LIES          ON
OPENCV_CONFIG_FILE_INCLUDE_DIR      /Users/Sysop/OpenCV-2.2.0/build
OPENCV_EXTRA_C_FLAGS
OPENCV_EXTRA_C_FLAGS_DEBUG
OPENCV_EXTRA_C_FLAGS_RELEASE
OPENCV_EXTRA_EXE_LINKER_FLAGS
OPENCV_EXTRA_EXE_LINKER_FLAGS_
OPENCV_EXTRA_EXE_LINKER_FLAGS_
OPENCV_WARNINGS_ARE_ERRORS          OFF
OPENEXR_INCLUDE_PATH                OPENEXR_INCLUDE_PATH-NOTFOUND
PVAPI_INCLUDE_PATH                  PVAPI_INCLUDE_PATH-NOTFOUND
TBB_INCLUDE_DIR                     /Users/Sysop/tbb30/include
TBB_LIB_DIR                         /Users/Sysop/tbb30/lib
USE_FAST_MATH                       ON
USE_IPP                             OFF
USE_O3                              ON
USE_OMIT_FRAME_POINTER              ON
WITH_1394                           ON
WITH_CARBON                         OFF
WITH_CUDA                           OFF
WITH_EIGEN2                         ON
WITH_FFMPEG                         ON
WITH_JASPER                         ON
WITH_JPEG                           ON
WITH_OPENEXR                        ON
WITH_PNG                            ON
WITH_PVAPI                          ON
WITH_QT                             OFF
WITH_QT_OPENGL                      OFF
WITH_QUICKTIME                      OFF

BUILD_DOXYGEN_DOCS: Generate HTML docs using Doxygen
Press [enter] to edit option
Press [c] to configure
Press [h] for help          Press [q] to quit without generating
Press [t] to toggle advanced mode (Currently Off)
```

Figure 3-3. ccmake for OpenCV in Mac OS X

Installing the Point Cloud Library (PCL)

What OpenCV is to computer vision, PCL is to 3-D data. PCL is a set of algorithms designed to help work with 3-D data, in particular point clouds. This library will be of great help when we work with point clouds but also when we work with voxels or even 3-D data in general. Again, you can be assured that if we use an algorithm from PCL, we'll be sure to explain it so that you're never in the dark and can return to the library again and again.

Windows

Installation of PCL is at about the same level of difficulty as installing OpenCV. The Windows installation is significantly more difficult, sadly, so we must download, make, and install many packages.

1. Download the latest stable release of Eigen (http://eigen.tuxfamily.org), FLANN (www.cs.ubc.ca/~mariusm/index.php/FLANN/FLANN), CMinPack (http://devernay.free.fr/hacks/cminpack/cminpack.html), VTKin the .exe installer form (www.vtk.org/VTK/resources/software.html), and Qhull (www.qhull.org/download/).

2. Unzip Eigen into a directory of your choice (e.g., C:\Eigen).

3. Copy the Eigen directory out of your C:\Eigen directory into the MinGW include directory (C:\MinGW\include\).

4. Unzip FLANN into a directory of your choice (e.g., C:\flann-x.y.z-src). Create a new folder in the directory called build.

5. Open cmake-gui, and set up the options as before, that is, set the source directory to be C:\<FLANN DIR>and the build directory to be C:\<FLANN DIR>\build.

6. Click Configure, and set the makefiles to MinGW. Use default native compilers. Click Configure again. Open the BUILD tab, and turn off both MatLab and Python bindings (unless you want them). Configure again, and your installation should look something like Figure 3-4. Generate and close the CMake GUI.

7. Open your command prompt. Navigate to C:\<FLANN DIR>\build, and run mingw32-make. After it finishes, open a command prompt as Administrator; navigate to C:\<FLANN DIR>\build, and run mingw32-make install.

Figure 3-4.cmake-guifor FLANN on Windows

8. Uncompress the tar file for CMinPack into a directory of your choice (e.g., C:\cminpack-1.1.3). Create a build directory in the CMinPack directory.

9. Run cmake-gui, and set the source code to the CMinPack directory and the build directory to the new folder you created in step 8.

10. Click Configure twice, settingthe options as shown in Figure 3-5, and thenclick Configure again. Generate and close cmake-gui.

11. Open a command prompt, and navigate to the CMinPack build directory. Then, run mingw32-make.Open a command prompt as Administrator; navigate to the same directory, and run mingw32-make install.

Figure 3-5. *cmake-gui for CMinPack on Windows*

12. Uncompress the tar file for Qhullinto a directory of your choice (e.g., `C:\qhull-2011.1`). Create a build directory in the Qhulldirectory.

13. Run `cmake-gui`, and set the source code to the Qhulldirectory and the build directory to the new folder you created in step 12.

14. Click Configure twice, setting the options as shown in Figure 3-6, and then click Configure again. Generate and close `cmake-gui`.

15. Open a command prompt; navigate over to the Qhullbuild directory, and then run `mingw32-make`. Open a command prompt as `Administrator`, navigate to the same directory, and run `mingw32-make install`.

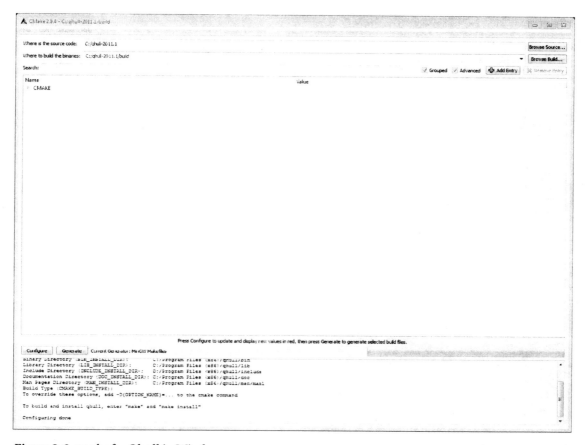

Figure 3-6.cmake for Qhull in Windows

16. Run the VTK installer.

17. Uncompress the tar file for PCL into a directory of your choice (e.g., C:\PCL_Install). Create a build directory in the PCL directory.

18. Run cmake-gui, and set the source code to the PCL directory and the build directory to the new folder you created in step 17.

19. Click Configure twice, setting the options as shown in Figure 3-7, and then click Configure again. Generate and close cmake-gui.

20. Open a command prompt; navigate to the PCL build directory, and run mingw32-make. Open a command prompt as Administrator; navigate to the same directory, and run mingw32-make install.

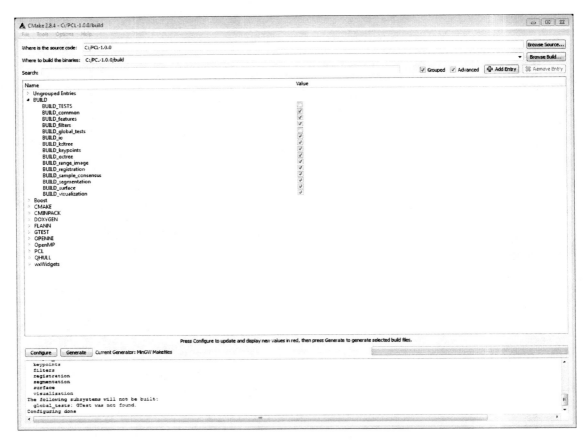

Figure 3-7.cmake for PCL in Windows

Linux

Both Ubuntu and Debian have supported distributions of PCL. Installation is a snap when running either of those distributions.

Ubuntu

For each of the newest four versions of Ubuntu, there's a personal package archive orPPA, that will automatically install PCL. Just run the following commands in your terminal:

```
sudo add-apt-repository ppa:v-launchpad-jochen-sprickerhof-de/pcl
sudo apt-get update
sudo apt-get install libpcl-all
```

Debian

Stable Debian systems are also supported, in particular, Maverick. Run the following commands to install on Maverick:

```
sudo apt-key adv --keyserver keyserver.ubuntu.com --recv-key 19274DEF
sudo echo/
"deb http://ppa.launchpad.net/v-launchpad-jochen-sprickerhof-de/pcl/ubuntu maverick main"/
>> /etc/apt/sources.list
sudo apt-get update
sudo apt-get install libpcl-all
```

You can install under an unstable system by replacing the references to maverickin the preceding commands with oneiric, but your installation will be unsupported by the package maintainer.

Mac OS X

We're going to be using MacPorts and installing PCL from source \ to get the most recent version installed. All of the code will be installed as Universal or 32-bit binaries. Here are the steps to follow:

1. Run the following MacPorts commands:

```
sudo port install boost +universal
sudo port install eigen3 +universal
sudo port install flann +universal
sudo port install vtk5 +universal+x11
sudo port install qhull +universal
sudo port install wxWidgets
```

2. At http://devernay.free.fr/hacks/cminpack/cminpack.html, download the most recent version of CMinPack.Uncompress the tar file. Create a build directory in the resultant folder.Run ccmake .. in the build folder. See Figure 3-8 for the proper CMake options. Configure, generate, and run make, and then run sudo make install.

3. Download the newest version of PCL from www.pointclouds.org/downloads/, and uncompress the tar file. Create another directory at the same level as the untarred file without the source (~/PCL-1.1.0/). Inside the directory you just created, run ccmake ../PCL-1.1.0-Source; see Figure 3-9 for the proper CMake options. Be certain to set the CMAKE_CXX_FLAGS and CMAKE_C_FLAGS to -m32.Configure, generate, and run make, and then run sudo make install.

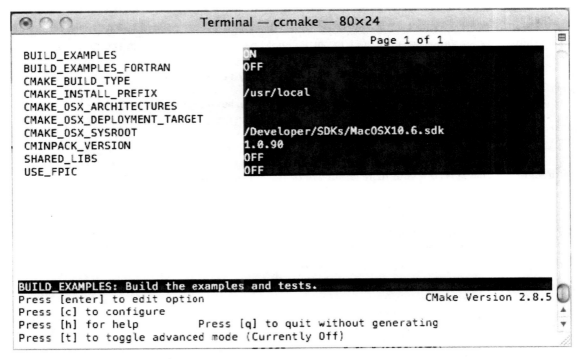

Figure 3-8.cmake options for CMinPack installation on Mac OS X

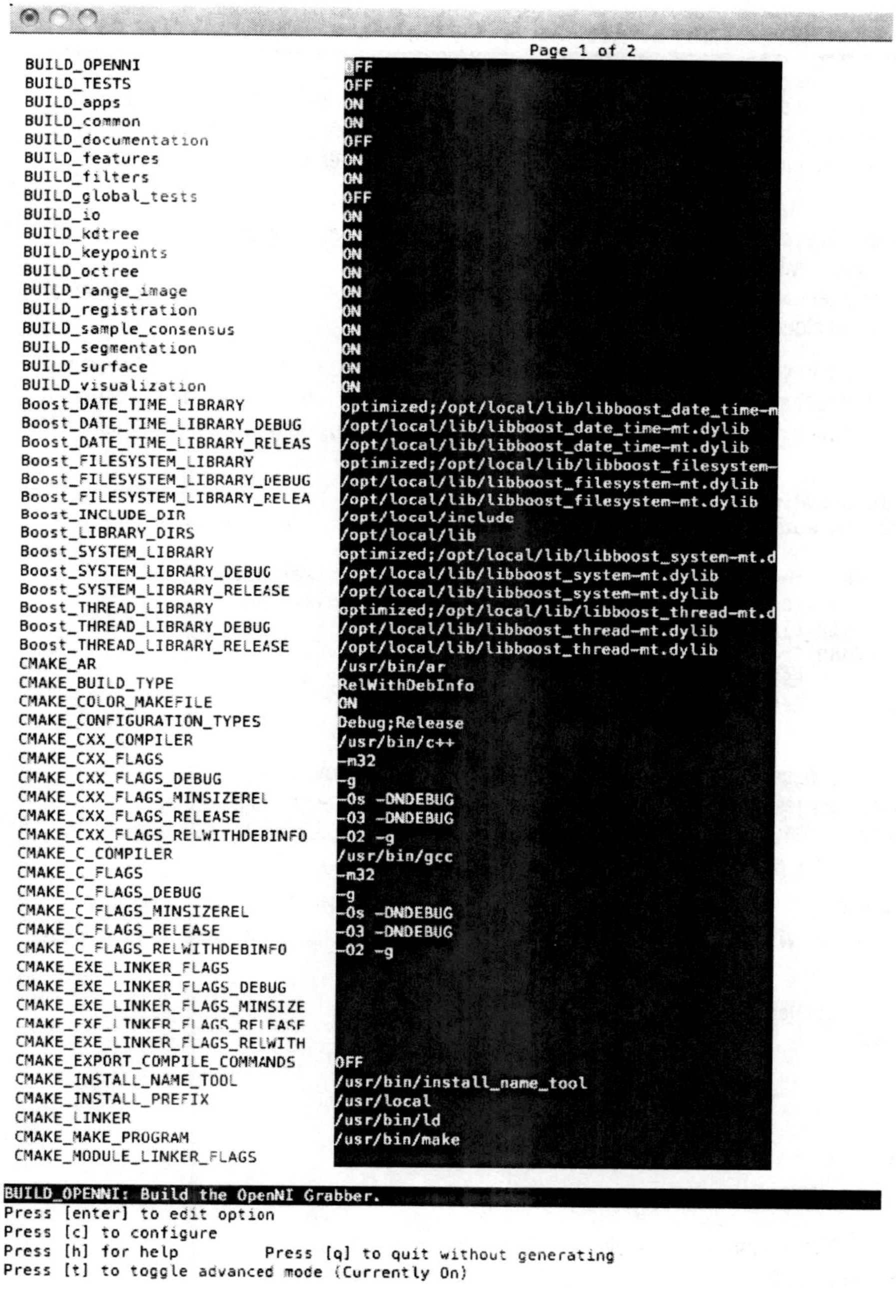

```
BUILD_OPENNI                         OFF
BUILD_TESTS                          OFF
BUILD_apps                           ON
BUILD_common                         ON
BUILD_documentation                  OFF
BUILD_features                       ON
BUILD_filters                        ON
BUILD_global_tests                   OFF
BUILD_io                             ON
BUILD_kdtree                         ON
BUILD_keypoints                      ON
BUILD_octree                         ON
BUILD_range_image                    ON
BUILD_registration                   ON
BUILD_sample_consensus               ON
BUILD_segmentation                   ON
BUILD_surface                        ON
BUILD_visualization                  ON
Boost_DATE_TIME_LIBRARY              optimized;/opt/local/lib/libboost_date_time-m
Boost_DATE_TIME_LIBRARY_DEBUG        /opt/local/lib/libboost_date_time-mt.dylib
Boost_DATE_TIME_LIBRARY_RELEAS       /opt/local/lib/libboost_date_time-mt.dylib
Boost_FILESYSTEM_LIBRARY             optimized;/opt/local/lib/libboost_filesystem-
Boost_FILESYSTEM_LIBRARY_DEBUG       /opt/local/lib/libboost_filesystem-mt.dylib
Boost_FILESYSTEM_LIBRARY_RELEA       /opt/local/lib/libboost_filesystem-mt.dylib
Boost_INCLUDE_DIR                    /opt/local/include
Boost_LIBRARY_DIRS                   /opt/local/lib
Boost_SYSTEM_LIBRARY                 optimized;/opt/local/lib/libboost_system-mt.d
Boost_SYSTEM_LIBRARY_DEBUG           /opt/local/lib/libboost_system-mt.dylib
Boost_SYSTEM_LIBRARY_RELEASE         /opt/local/lib/libboost_system-mt.dylib
Boost_THREAD_LIBRARY                 optimized;/opt/local/lib/libboost_thread-mt.d
Boost_THREAD_LIBRARY_DEBUG           /opt/local/lib/libboost_thread-mt.dylib
Boost_THREAD_LIBRARY_RELEASE         /opt/local/lib/libboost_thread-mt.dylib
CMAKE_AR                             /usr/bin/ar
CMAKE_BUILD_TYPE                     RelWithDebInfo
CMAKE_COLOR_MAKEFILE                 ON
CMAKE_CONFIGURATION_TYPES            Debug;Release
CMAKE_CXX_COMPILER                   /usr/bin/c++
CMAKE_CXX_FLAGS                      -m32
CMAKE_CXX_FLAGS_DEBUG                -g
CMAKE_CXX_FLAGS_MINSIZEREL           -Os -DNDEBUG
CMAKE_CXX_FLAGS_RELEASE              -O3 -DNDEBUG
CMAKE_CXX_FLAGS_RELWITHDEBINFO       -O2 -g
CMAKE_C_COMPILER                     /usr/bin/gcc
CMAKE_C_FLAGS                        -m32
CMAKE_C_FLAGS_DEBUG                  -g
CMAKE_C_FLAGS_MINSIZEREL             -Os -DNDEBUG
CMAKE_C_FLAGS_RELEASE                -O3 -DNDEBUG
CMAKE_C_FLAGS_RELWITHDEBINFO         -O2 -g
CMAKE_EXE_LINKER_FLAGS
CMAKE_EXE_LINKER_FLAGS_DEBUG
CMAKE_EXE_LINKER_FLAGS_MINSIZE
CMAKE_EXE_LINKER_FLAGS_RELEASE
CMAKE_EXE_LINKER_FLAGS_RELWITH
CMAKE_EXPORT_COMPILE_COMMANDS        OFF
CMAKE_INSTALL_NAME_TOOL              /usr/bin/install_name_tool
CMAKE_INSTALL_PREFIX                 /usr/local
CMAKE_LINKER                         /usr/bin/ld
CMAKE_MAKE_PROGRAM                   /usr/bin/make
CMAKE_MODULE_LINKER_FLAGS

BUILD_OPENNI: Build the OpenNI Grabber.
Press [enter] to edit option
Press [c] to configure
Press [h] for help          Press [q] to quit without generating
Press [t] to toggle advanced mode (Currently On)
```

Figure 3-9.cmakeoptions for PCL in Mac OS X

GETTING OPENKINECT INTO PCL

In this exercise, we're going to create a bridge between the OpenKinect driver and PCL to quickly and efficiently move data captured by the Kinect into PCL.PCL has an OpenNI driver system that handles this for you, but because we're going with a purely open source solution, this driver isn't available to us. Let's get started.

The OpenKinect code in this system is derived from the `cppview` demonstration program included with your OpenKinect installation. We're going to be using a block of that code, cutting out the OpenGL requirements, and bolting on the PCL system. We're also going to alter the some of the OpenKinect internal code to create proper point cloud output.

First, create a new C++ file in your directory—we used `OKPCL.cpp`—and open it in your editor of choice. Then, in the OpenKinect directory, under `libfreenect/src`, open `registration.c`. Next, find the code listing that starts with `///camera -> world coordinate helper function`, and replace it with the following:

```
/// camera -> world coordinate helper function
voidfreenect_camera_to_world(freenect_device*dev,intcx,intcy,intwz,double*wx,double*wy)
{
doubleref_pix_size=dev->registration.zero_plane_info.reference_pixel_size;
doubleref_distance=dev->registration.zero_plane_info.reference_distance;
doublexfactor=2*ref_pix_size*wz/ref_distance;
doubleyfactor=(1024/480)*ref_pix_size*wz/ref_distance;
*wx=(double)(cx-DEPTH_X_RES/2)*xfactor;
*wy=(double)(cy-DEPTH_Y_RES/2)*yfactor;
}
```

We are changing this code, because the Kinect's image sensor performs auneven compression of the collected data when it packages it to be sent to the computer. While the X direction is a perfect factor of two (1280 [input] divided by 640 [output]), the Y direction is not (1024 [input] divided by 480 [output]). We correct for these factors in the preceding code.

It's time for another edit.Open `libfreenect/wrappers/cpp/libfreenect.hpp`. At the top, replace `#include <libfreenect.h>` with `#include <libfreenect-registration.h>`. Then, under the following code,

```
freenect_depth_formatgetDepthFormat(){
returnm_depth_format;
}
```

addthis new method:

```
freenect_device*getDevice(){
returnm_dev;
}
```

These changes bring registration into the C++ side of things, and let us access the device pointer from outside of the device itself (dangerous, but necessary). Now that you've made these edits, return to your build directory (`libfreenect/build`) and recompile and reinstall as you did when you first installed libfreenect. This will give you access to all of these features in your other code. Be sure to write down or remember where these files are installed! You're going to need to use them later with your new code.

Now, let's return to our new file, OKPCL.cpp. We're going to start by laying out some boilerplate code, taken straight from the cppview demonstration. Some parts have been deleted and some added, but it is mostly the same. This code will handle all of the communication underlying between the raw driver and our higher level system:

```
/*
 * This file is part of the OpenKinect Project. http://www.openkinect.org
 *
 * Copyright (c) 2010 individual OpenKinect contributors. See the CONTRIB file
 * for details.
 *
 * This code is licensed to you under the terms of the Apache License, version
 * 2.0, or, at your option, the terms of the GNU General Public License,
 * version 2.0. See the APACHE20 and GPL2 files for the text of the licenses,
 * or the following URLs:
 * http://www.apache.org/licenses/LICENSE-2.0
 * http://www.gnu.org/licenses/gpl-2.0.txt
 *
 * If you redistribute this file in source form, modified or unmodified, you
 * may:
 *    1) Leave this header intact and distribute it under the same terms,
 *       accompanying it with the APACHE20 and GPL20 files, or
 *    2) Delete the Apache 2.0 clause and accompany it with the GPL2 file, or
 *    3) Delete the GPL v2 clause and accompany it with the APACHE20 file
 * In all cases you must keep the copyright notice intact and include a copy
 * of the CONTRIB file.
 *
 * Binary distributions must follow the binary distribution requirements of
 * either License.
 */

#include<iostream>
#include<libfreenect.hpp>
#include<pthread.h>
#include<stdio.h>
#include<string.h>
#include<cmath>
#include<vector>
#include<ctime>
#include<boost/thread/thread.hpp>
#include"pcl/common/common_headers.h"
#include"pcl/io/pcd_io.h"
#include"pcl/visualization/pcl_visualizer.h"
#include"pcl/console/parse.h"
#include"pcl/point_types.h"
#include"boost/lexical_cast.hpp"
```

```
///Mutex Class

class Mutex {
public:
Mutex() {
pthread_mutex_init( &m_mutex, NULL );
}
void lock() {
pthread_mutex_lock( &m_mutex );
}
void unlock() {
pthread_mutex_unlock( &m_mutex );
}

class ScopedLock
{
Mutex & _mutex;
public:
ScopedLock(Mutex & mutex)
: _mutex(mutex)
{
_mutex.lock();
}
~ScopedLock()
{
_mutex.unlock();
}
};
private:
pthread_mutex_t m_mutex;
};


///Kinect Hardware Connection Class

/* thanks to Yoda---- from IRC */
class MyFreenectDevice : public Freenect::FreenectDevice {
public:
MyFreenectDevice(freenect_context *_ctx, int _index)
: Freenect::FreenectDevice(_ctx, _index),
depth(freenect_find_depth_mode(FREENECT_RESOLUTION_MEDIUM,
FREENECT_DEPTH_REGISTERED).bytes),
m_buffer_video(freenect_find_video_mode(FREENECT_RESOLUTION_MEDIUM,
FREENECT_VIDEO_RGB).bytes), m_new_rgb_frame(false), m_new_depth_frame(false)
{

}
//~MyFreenectDevice(){}

// Do not call directly even in child
void VideoCallback(void* _rgb, uint32_t timestamp) {
Mutex::ScopedLock lock(m_rgb_mutex);
uint8_t* rgb = static_cast<uint8_t*>(_rgb);
```

```
std::copy(rgb, rgb+getVideoBufferSize(), m_buffer_video.begin());
m_new_rgb_frame = true;
};
// Do not call directly even in child
void DepthCallback(void* _depth, uint32_t timestamp) {
Mutex::ScopedLock lock(m_depth_mutex);
depth.clear();
uint16_t* call_depth = static_cast<uint16_t*>(_depth);
for (size_t i = 0; i < 640*480 ; i++) {
depth.push_back(call_depth[i]);
}
m_new_depth_frame = true;
}
bool getRGB(std::vector<uint8_t>&buffer) {
//printf("Getting RGB!\n");
Mutex::ScopedLock lock(m_rgb_mutex);
if (!m_new_rgb_frame) {
//printf("No new RGB Frame.\n");
returnfalse;
}
buffer.swap(m_buffer_video);
m_new_rgb_frame = false;
returntrue;
}

bool getDepth(std::vector<uint16_t>&buffer) {
Mutex::ScopedLock lock(m_depth_mutex);
if (!m_new_depth_frame)
returnfalse;
buffer.swap(depth);
m_new_depth_frame = false;
returntrue;
}

private:
std::vector<uint16_t> depth;
std::vector<uint8_t> m_buffer_video;
Mutex m_rgb_mutex;
Mutex m_depth_mutex;
bool m_new_rgb_frame;
bool m_new_depth_frame;
};
```

Next, we're going to add our first part of the bridging code. This will create all the necessary variables for our code.

```
///Start the PCL/OK Bridging

//OK
Freenect::Freenect freenect;
MyFreenectDevice* device;
freenect_video_format requested_format(FREENECT_VIDEO_RGB);
```

```
double freenect_angle(0);
int got_frames(0),window(0);
int g_argc;
char **g_argv;
int user_data = 0;


//PCL
pcl::PointCloud<pcl::PointXYZRGB>::Ptr cloud (new pcl::PointCloud<pcl::PointXYZRGB>);
unsignedint cloud_id = 0;
```

With that added, we're ready to put in our main function. Let's go over it one piece at a time. This first part will create the two empty data structures to hold our raw data, as well as the PCL point cloud we're going to display. We will then set up the viewer that will do the actual hard work of displaying our cloud on the screen. You may note the registration of a keyboard callback function; we don't have that in yet, but we'll add it shortly!

```
// --------------
// -----Main-----
// --------------
int main (int argc, char** argv)
{
//More Kinect Setup

static std::vector<uint16_t> mdepth(640*480);
static std::vector<uint8_t> mrgb(640*480*4);

// Fill in the cloud data
cloud->width    = 640;
cloud->height   = 480;
cloud->is_dense = false;
cloud->points.resize (cloud->width * cloud->height);

// Create and setup the viewer


boost::shared_ptr<pcl::visualization::PCLVisualizer> viewer (new
pcl::visualization::PCLVisualizer ("3D Viewer"));
viewer->registerKeyboardCallback (keyboardEventOccurred, (void*)&viewer);
viewer->setBackgroundColor (255, 255, 255);
viewer->addPointCloud<pcl::PointXYZRGB> (cloud, "Kinect Cloud");
viewer->setPointCloudRenderingProperties (pcl::visualization::PCL_VISUALIZER_POINT_SIZE,
1, "Kinect Cloud");
viewer->addCoordinateSystem (1.0);
viewer->initCameraParameters ();
```

Great! Now let's start up the Kinect itself. We're going to start the device, then grab depth data from it until we get a clean response, that is, until all depth voxels are reporting in.

```
device = &freenect.createDevice<MyFreenectDevice>(0);
device->startVideo();
device->startDepth();
boost::this_thread::sleep (boost::posix_time::seconds (1));
```

```
//Grab until clean returns
int DepthCount = 0;
while (DepthCount == 0) {
device->updateState();
device->getDepth(mdepth);
device->getRGB(mrgb);
for (size_t i = 0;i < 480*640;i++)
DepthCount+=mdepth[i];
}

device->setVideoFormat(requested_format);
```

Finally, we come to the main loop. Here, we're going to grab the data from the Kinect, and then read it into our point cloud. Next, we're going to display that point cloud in the viewer.We'll do this until the viewer is stopped (accomplished by pressing the Q key in the viewer window).

```
//-------------------
// -----Main loop-----
//-------------------
double x = NULL;
double y = NULL;
int iRealDepth = 0;
while (!viewer->wasStopped ())
{
device->updateState();
device->getDepth(mdepth);
device->getRGB(mrgb);

size_t i = 0;
size_t cinput = 0;
for (size_t v=0 ; v<480 ; v++)
{
for ( size_t u=0 ; u<640 ; u++, i++)
{
iRealDepth = mdepth[i];
freenect_camera_to_world(device->getDevice(), u, v, iRealDepth, &x, &y);
cloud->points[i].x  = x;
cloud->points[i].y  = y;
cloud->points[i].z = iRealDepth;
cloud->points[i].r = mrgb[i*3];
cloud->points[i].g = mrgb[(i*3)+1];
cloud->points[i].b = mrgb[(i*3)+2];
}
}

viewer->updatePointCloud (cloud, "Kinect Cloud");
viewer->spinOnce ();
}
device->stopVideo();
device->stopDepth();
return 0;
}
```

So, beyond the Q key (which quits), there's also the R key functionality, which recenters the view so you can see all of the cloud. However, we want to add one more thing:using theC key to capture a point cloud and write it out to disk so you can look at it later or use it with other software. The following code adds that functionality:

```
///Keyboard Event Tracking
void keyboardEventOccurred (const pcl::visualization::KeyboardEvent &event,
void* viewer_void)
{
  boost::shared_ptr<pcl::visualization::PCLVisualizer> viewer =
*static_cast<boost::shared_ptr<pcl::visualization::PCLVisualizer> *> (viewer_void);
if (event.getKeySym () == "c"&&event.keyDown ())
  {
    std::cout <<"c was pressed => capturing a pointcloud"<< std::endl;
    std::string filename = "KinectCap";
    filename.append(boost::lexical_cast<std::string>(cloud_id));
    filename.append(".pcd");
    pcl::io::savePCDFileASCII (filename, *cloud);
    cloud_id++;
  }
}
```

Now, you'll need to create another file in the directory—CMakeLists.txt—and add the following code to it. This file tells the CMake system how to compile your new code. If you used a name for your file other than OKPCL.cpp, be sure to replace that in the following code.

```
cmake_minimum_required(VERSION 2.8 FATAL_ERROR)
set(CMAKE_C_FLAGS "-Wall")

project(OKPCL)

if (WIN32)
set(THREADS_USE_PTHREADS_WIN32 true)
find_package(Threads REQUIRED)

include_directories(${THREADS_PTHREADS_INCLUDE_DIR})
endif()

find_package(PCL 1.4 REQUIRED)

include_directories(. ${PCL_INCLUDE_DIRS})
link_directories(${PCL_LIBRARY_DIRS})
add_definitions(${PCL_DEFINITIONS})

add_executable(OKPCL OKPCL.cpp)

if(APPLE)
target_link_libraries(OKPCL freenect ${PCL_LIBRARIES})
else()
find_package(Threads REQUIRED)
include_directories(${USB_INCLUDE_DIRS})
```

```
if (WIN32)
set(MATH_LIB "")
else(WIN32)
set(MATH_LIB "m")
endif()
target_link_libraries(OKPCL freenect ${CMAKE_THREAD_LIBS_INIT} ${MATH_LIB}
${PCL_LIBRARIES})
endif()
```

Save that file, and exit your editor. Recall that we said earlier that you needed to remember where your installer put the libfreenect files? You'll need them now.Copy libfreenect.hpp, libfreenect-registration.h, and libfreenect.h into your OKPCL directory. Run ccmakeand make as you did in all of the previous installations we've performed. Your code should compile, and you're ready to go! Plug in your Kinect, run your code, and spin around the point cloud output.

Summary

In this chapter, we installed two very powerful software packages—PCL and OpenCV—and showed you how to link your installation of OpenKinect (the only open source Kinect driver) to PCL. You should now be able to generate point clouds in the viewer and capture them for viewing later.

Computer Vision

Several powerful image processing techniques can be made even more effective when used with a depth image from a Kinect. In this chapter, we will go over how to perform basic image processing. We will then discuss how these techniques work with standard camera input, what they are used for, and how the Kinect's depth image can be used in place of the normal image for different results or to solve some of the issues that can occur with these traditional approaches.

Anatomy of an Image

Digital images are made of pixels. A pixel is the single smallest unit of an image. Images have rows and columns of pixels; the number of columns by the number of rows is said to be the image's resolution. The more pixels an image has, the more faithful a representation can be made. Then too, the more pixels an image has, the more processing is required to analyze it, which can slow down your program. In general, for image processing, it is ideal to keep the image as small as possible while maintaining the desired fidelity. Figure 4-1 shows some different resolution examples.

Figure 4-1. Icons (left) can be as small as 8 × 8 pixels (64 total pixels). A 1024 × 728 resolution image (right) has 745,472 pixels. The Kinect RGB and depth cameras (center) are both 640 × 480 (307,200 pixels).

A color image will be made up of color pixels, while a greyscale image will be made up of various shades of gray between the two extremes of black and white. Color pixels are made of different amounts of red, green, and blue (RGB), while greyscale images are a single value representing how bright, or how white, that pixel is. The Kinect provides a color image from its RGB camera and greyscale image from its depth camera. Both images have a resolution of 640 × 480.

Image Processing Basics

Image processing encompasses a number of techniques to make an image more amenable to interpretation, manipulation, and display. You can simplify an image by, for example, removing color. You can apply techniques to reduce the effects from image *noise*, which is unwanted fluctuation produced by the electronics in digital cameras. The next few sections go over some fundamental image processing techiques.

Simplifying Data

Many image processing techniques focus on simplifying data, literally changing a detailed image (either greyscale or color) into a simpler black-and-white image, where the white portions of the image indicate areas of interest (for example, the position of an object). These images can be combined with other techniques or used individually. Figure 4-2 shows a greyscale image that has been simplified into a black-and-white image in order to highlight the area of focus.

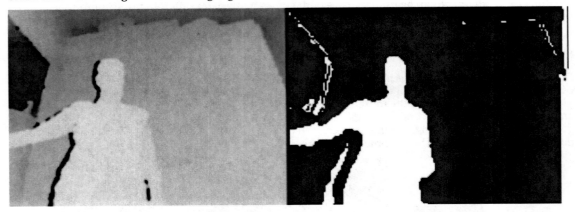

Figure 4-2. A greyscale image from the camera (left) and a black-and-white image that has been simplified by image processing to highlight the focus(right)

Many image libraries, including the ones used in this book, use a one-dimensional array to store all the pixels of an image. This makes image processing faster but also requires a little more work to get a pixel from a specific x and y location.

Let's say that you had a very small image, 5x5 pixels. That is a total of 25 pixels. The first 5 pixels in the array make up the first line of the image, thus positions 0-4 in the array represent the top row of pixels in the image. (Remember, arrays start at position 0.) The 6th pixel in the array, in position 5, wraps around to the next row of the image. Take a look at the Figure 4-3

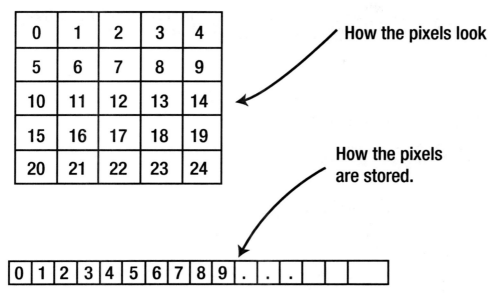

Figure 4-3. The mapping of a pixel grid to a pixel array

So, if we wanted to know the value of the pixel when *x* and *y* equal 1, we would look at the value of position 6 in the array (the rows and columns start at 0, just like the array). You can use the following formula to determine which array you want from a given value of *x* and *y*:

```
int index =  x + y * width;
```

where `index` is the position in the array you want to access and `width` represents the width of the image.

■ **Note** You may want to reverse the image from the camera. This is useful if you want to make the camera input work like a mirror. To do this, you need only to alter the formula to

```
int index =  (width - x - 1) + y * width;
```

Noise and Blurring

Digital cameras produce images that are subject to image noise, which is a variation in brightness or color of pixels that is not present in reality. Higher quality cameras will produce less noise, but it will still occur. Both the RGB and depth image of the Kinect are very noisy.

We can employ processes to reduce the effect of noise. One such approach is to blur the image. There are many different blurring techniques, but some can be slow and thus not ideal for real-time image processing. The function in Listing 4-1 shows a box blur, a quick and simple blurring method, good for real time image processing. Instead of returning the value of the pixel, Box Blurring returns the average value of the pixel and its surrounding pixels specified by parameters passed to the function.

Listing 4-1. *Blurring Algorithm*

```
#include "testApp.h"
float testApp::blur(ofImage* img, int x, int y, int blurSize){
        float greyLevel = 0;
        unsigned char* pixels = img->getPixels();
        int numPixels = 0;
    for(int dx = -blurSize; dx <= blurSize; dx++){
                for(int dy = -blurSize; dy <= blurSize; dy++){


                        int newX = ofClamp((dx + x), 0, greyImage.getWidth()  - 1);
                        int newY = ofClamp((dy + y), 0, greyImage.getHeight() - 1);
                        numPixels++;
                        int i =  (newX + newY * img->getWidth());
                        greyLevel += pixels[i];


                }
    }

    greyLevel = greyLevel/numPixels;
        return greyLevel;
}
```

Figure 4-4 shows the result of an image blurred with the algorithm in Listing 4-1. You can see the effects from different blurring boxes.

Figure 4-4. *The original image (right), the image blurred with a box of two neighboring pixels (center), and the image blurred with a box of four neighboring pixels (right)*

The more an image is blurred, the less the image will look like real life. What's more, the larger the blur size, the slower the program will run. The extent to which blurring is required depends on how much noise your image contains. We recommend starting with a blur size of at least 2.

Image blurring is not only effective for removing noise, but is also helpful when comparing images, as you will see later in this chapter.

Contriving Your Situation

Even after employing techniques like blurring and thresholding (which is discussed in the next section), there will most likely still be some issues with your image processing. Computer vision is an advanced field in computer science and even very complicated algorithms cannot ensure accuracy for all situations.

It is far simpler, therefore, to contrive your situation to be a controlled environment that encourages consistent and predictable results. It is easier to write a program that can distinguish an object from a flat, solid-color background than from an arbitrary background of colors and shapes. The upcoming technique of brightness thresholding works better when you can control for the background. For example, you will notice our images in this chapter feature a white wall with a single person in front of it. You may not be able to duplicate this situation exactly, but the simpler you can make your scene, the better your results will be.

Brightness Thresholding

Now that you know how pixels are arranged, how to blur an image, and how to contrive your situation, let's consider a simple example that will create a black-and-white image. Listing 4-2 uses a technique called *brightness thresholding*, where any pixel at or below a certain brightness level will set to black, and any pixel above that level will be set to white.

Listing 4-2. Brightness thresholding algorithm

```
#include "testApp.h"
int mod = 4;
float threshold = 150;

//----------------------------------------------------------------
void testApp::setup() {
        kinect.init();
        kinect.setVerbose(true);
        kinect.open();

        resultImage.allocate(kinect.width/mod, kinect.height/mod, OF_IMAGE_GRAYSCALE);
            ofSetFrameRate(60);
        // zero the tilt on startup
        angle = 0;
        kinect.setCameraTiltAngle(angle);
}

//----------------------------------------------------------------
void testApp::update() {
        ofBackground(100, 100, 100);
        threshold = ofMap(mouseX, 0, ofGetViewportWidth(), 0, 255, 255);

        kinect.update();
        if(kinect.isFrameNew())    // there is a new frame and we are connected
        {
```

```
                    greyImage.setFromPixels(kinect.getPixels(), kinect.getWidth(),
        kinect.getHeight(), OF_IMAGE_COLOR,true);
                    //greyImage.setFromPixels(kinect.getDepthPixels(), kinect.getWidth(),
        kinect.getHeight(), OF_IMAGE_GRAYSCALE,true);
                    greyImage.setImageType(OF_IMAGE_GRAYSCALE);
                    greyImage.resize(greyImage.getWidth()/mod, greyImage.getHeight()/mod);

                    unsigned char * pixels = resultImage.getPixels();

                    for(int x = 0; x < greyImage.width; x++){
                            for(int y = 0; y < greyImage.height; y++){
                                    int i = x + y * greyImage.width;

                                    int color = blur(&greyImage, x, y, 1);

                                    pixels[i] = color;

                                    if(color > threshold){
                                            pixels[i] = 255;
                                    } else {
                                            pixels[i] = 0;
                                    }
                            }
                    }

                    resultImage.update();
            }
    }

    //----------------------------------------------------------------
    void testApp::draw() {
            ofSetColor(255, 255, 255);
            resultImage.draw(0, 0, 640, 480);
    }


    //----------------------------------------------------------------
    void testApp::exit() {
            kinect.setCameraTiltAngle(0); // zero the tilt on exit
            kinect.close();
    }


    //----------------------------------------------------------------
    float testApp::blur(ofImage* img, int x, int y, int blurSize){
            float greyLevel = 0;

            unsigned char* pixels = img->getPixels();

            int numPixels = 0;

        for(int dx = -blurSize; dx <= blurSize; dx++){
```

```
                for(int dy = -blurSize; dy <= blurSize; dy++){

                        int newX = ofClamp((dx + x), 0, greyImage.getWidth()  - 1);
                        int newY = ofClamp((dy + y), 0, greyImage.getHeight() - 1);

                        numPixels++;

                        int i =  (newX + newY * img->getWidth());

                        greyLevel += pixels[i];

                }
        }

    greyLevel = greyLevel/numPixels;

    return greyLevel;
}
//--------------------------------------------------------------
void testApp::keyPressed (int key) {
        switch (key) {
                case OF_KEY_UP:
                        angle++;
                        if(angle>30) angle=30;
                        kinect.setCameraTiltAngle(angle);
                        break;

                case OF_KEY_DOWN:
                        angle--;
                        if(angle<-30) angle=-30;
                        kinect.setCameraTiltAngle(angle);
                        break;
        }
}
```

The preceding example uses the Kinect's color image, but the result is quite different if we change it to use the depth image instead, by taking the following steps:

1. Comment out this line:

```
greyImage.setFromPixels(kinect.getPixels(), kinect.getWidth(), kinect.getHeight(),
OF_IMAGE_COLOR,true);
```

2. And add in this line:

```
greyImage.setFromPixels(kinect.getDepthPixels(), kinect.getWidth(), kinect.getHeight(),
OF_IMAGE_GRAYSCALE,true);
```

The Kinect depth image shows parts of the image that are closer to the sensor as lighter and parts that are further away from the sensor as darker. When you apply the program in Listing 4-2, along with the preceding modification, to the depth image, the result actually shows items that are within a certain distance of the camera as white and anything further away as black.

Figure 4-5 shows the result from processing a Kinect depth image.

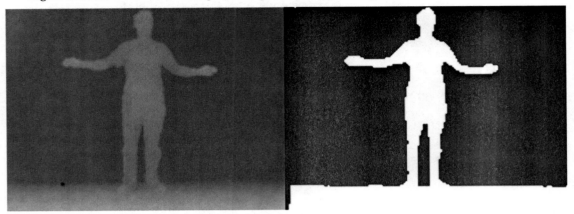

Figure 4-5. *Kinect brightness image*

Brightest Pixel Tracking

A simple interface can be created with a brightest pixel tracking algorithm. By giving the user a bright object, like a flashlight or an LED, you can track a user's movements by following the brightest pixel created by the light source, as shown in Listing 4-3.

Listing 4-3. Brightest Pixel Tracking Algorithm:

```
#include "testApp.h"
int mod = 4;
float threshold = 150;
int brightestX, brightestY;
//-----------------------------------------------------------------
void testApp::setup() {
        kinect.init();
        kinect.setVerbose(true);
        kinect.open();

        resultImage.allocate(kinect.width/mod, kinect.height/mod, OF_IMAGE_GRAYSCALE);
        ofSetFrameRate(60);
        // zero the tilt on startup
        angle = 0;
        kinect.setCameraTiltAngle(angle);
}

//-----------------------------------------------------------------
void testApp::update() {
        ofBackground(100, 100, 100);
        threshold = ofMap(mouseX, 0, ofGetViewportWidth(), 0, 255, 255);
```

```
                kinect.update();
                if(kinect.isFrameNew())   // there is a new frame and we are connected
                {

                        greyImage.setFromPixels(kinect.getPixels(), kinect.getWidth(),
kinect.getHeight(), OF_IMAGE_COLOR, true);
                        //greyImage.setFromPixels(kinect.getDepthPixels(), kinect.getWidth(),
kinect.getHeight(), OF_IMAGE_GRAYSCALE, true);
                        greyImage.setImageType(OF_IMAGE_GRAYSCALE);
                        greyImage.resize(greyImage.getWidth()/mod, greyImage.getHeight()/mod);
                        unsigned char * pixels = resultImage.getPixels();
                        int brightest = 0;
                        for(int x = 0; x < greyImage.width; x++){
                                for(int y = 0; y < greyImage.height; y++){
                                        int i = x + y * greyImage.width;
                                        int color = blur(&greyImage, x, y, 1);
                                        pixels[i] = color;
                                        if(color >= brightest){
                                                brightest = color;
                                                brightestX = x;
                                                brightestY = y;
                                                cout << brightestX << "x" << brightestY <<endl;
                                        }
                                }
                        }

                        resultImage.update();
                }
}

//----------------------------------------------------------------
void testApp::draw() {
        ofSetColor(255, 255, 255);
        resultImage.draw(0, 0, 640, 480);

        ofSetColor(255, 0, 0);
        ofEllipse(brightestX * mod, brightestY * mod, 50, 50);
}


//----------------------------------------------------------------
void testApp::exit() {
        kinect.setCameraTiltAngle(0); // zero the tilt on exit
        kinect.close();
}


//----------------------------------------------------------------
float testApp::blur(ofImage* img, int x, int y, int blurSize){
        float greyLevel = 0;
        unsigned char* pixels = img->getPixels();
        int numPixels = 0;
```

```
    for(int dx = -blurSize; dx <= blurSize; dx++){
            for(int dy = -blurSize; dy <= blurSize; dy++){

                    int newX = ofClamp((dx + x), 0, greyImage.getWidth()  - 1);
                    int newY = ofClamp((dy + y), 0, greyImage.getHeight() - 1);
                    numPixels++;
                    int i =  (newX + newY * img->getWidth());
                    greyLevel += pixels[i];

            }
    }

    greyLevel = greyLevel/numPixels;
        return greyLevel;
}

//----------------------------------------------------------------
void testApp::keyPressed (int key) {
        switch (key) {
                case OF_KEY_UP:
                        angle++;
                        if(angle>30) angle=30;
                        kinect.setCameraTiltAngle(angle);
                        break;

                case OF_KEY_DOWN:
                        angle--;
                        if(angle<-30) angle=-30;
                        kinect.setCameraTiltAngle(angle);
                        break;
        }
}
```

When used with the depth image of the Kinect, brightest pixel tracking has a completely different effect. The brightest pixel will be the pixel closest to the Kinect. Tracking that pixel is useful for simple interfaces that involve tracking an object—for example, an extended hand. The modification to the code in Listing 4-3 is simple. Simply comment out the following line:

```
greyImage.setFromPixels(kinect.getPixels(), kinect.getWidth(), kinect.getHeight(),
OF_IMAGE_COLOR,true);
```

And add in this line:

```
greyImage.setFromPixels(kinect.getDepthPixels(), kinect.getWidth(), kinect.getHeight(),
OF_IMAGE_GRAYSCALE,true);
```

Comparing Images

Computer programs can receive several images per second from a camera input. Each one of these images is called a *frame*. The next techniques we'll explore in this chapter will center around comparing the current frame's image with another frame's.

When comparing images, it is important to realize that even when the subject of a camera has not changed, the pixel values in images captured by that camera will seldom be identical.

Thresholding with a Tolerance

Even with blurring pixels, values will likely differ some between frames. To compensate for this, we introduce a tolerance variable to use when comparing images. Rather than comparing the different values directly, we check to see if the new value is within a certain range (tolerance) of the value it's being compared to. So, rather than

```
newPixel != oldPixel
```

we use

```
(newPixel + tolerance <= oldPixel) || (newPixel - tolerance >= oldPixel)
```

■ **Note** Adjusting the tolerance level appropriately for the level of noise present in your image is important. If the tolerance is too low, the image comparison will still suffer from issues resulting from noise. If the tolerance is too high, the comparison will ignore valid differences between images.

Figure 4-6 shows the effects from having your tolerance too low or too high.

Figure 4-6. *Tolerance too low (left), tolerance too high (center), and tolerance at the correct level (right)*

Now that we have established how to blur pixels and remove noise, we can compare two different images. We want to create a black-and-white image that shows the differences between one frame and another. To do so, we need to compare the pixels in each position of an image with the pixels in the same positions in another image. The algorithm is demonstrated in Listing 4-4.

Listing 4-4. *Algorithm for image comparison*

```
for(int x = 0; x < currentImage.width; x++){
        for(int y = 0; y < currentImage.height; y++){
                int i = x + y * currentImage.width;
                int color = blur(&currentImage, x, y, 10);
                int prevColor = prevImage.getPixels()[i];
```

```
                                    if((color + threshold > prevColor) || (prevColor < color +
    threshold)){
                                        pixels[i] = 0;
                            } else {
                                    pixels[i] = 255;
                            }
                        }
                }
```

Background Subtraction

One common algorithm for image processing is called *background subtraction*. Figure 4-7 shows an example. In background subtraction, the algorithm first stores a single frame from the camera and then uses it for a basis of comparison. The resulting image will show anything that has changed in the scene.

Figure 4-7. A background image (left), a new image (center), and the image resulting from comparison (right)

Once you understand how to compare images, background subtraction is a simple matter of deciding when to capture the scene. It is useful to attach the capturing of the user image to a user key press, as we do in Listing 4-5 (in this case, the return key).

Listing 4-5. Code for Capturing an Image on a Keypress (for Background Subtraction)

```
void testApp::keyPressed (int key) {
        switch (key) {
                case OF_KEY_RETURN:
                        prevImage.setFromPixels(
                    currentImage.getPixels(),
                    currentImage.getWidth(), currentImage.getHeight(), OF_IMAGE_GRAYSCALE, true);
                        break;
        }
}
```

The completed code for background subtraction is in Listing 4-6:

Listing 4-6. *Code for Background Subtraction*

```
#include "testApp.h"

int mod = 4;

float threshold = 150;

//-------------------------------------------------------------
void testApp::setup() {
    kinect.init();
    kinect.setVerbose(true);
    kinect.open();

    prevImage.allocate(kinect.width/mod, kinect.height/mod, OF_IMAGE_GRAYSCALE);
    resultImage.allocate(kinect.width/mod, kinect.height/mod, OF_IMAGE_GRAYSCALE);

    ofSetFrameRate(60);

    // zero the tilt on startup
    angle = 0;
    kinect.setCameraTiltAngle(angle);
}

//-------------------------------------------------------------
void testApp::update() {
    ofBackground(100, 100, 100);

    threshold = ofMap(mouseX, 0, ofGetViewportWidth(), 0, 255, 255);

    kinect.update();
    if(kinect.isFrameNew())     // there is a new frame and we are connected
    {

            currentImage.setFromPixels(kinect.getPixels(), kinect.getWidth(), kinect.getHeight(),
OF_IMAGE_COLOR, true);
            //currentImage.setFromPixels(kinect.getDepthPixels(), kinect.getWidth(),
kinect.getHeight(), OF_IMAGE_GRAYSCALE, true);
            currentImage.mirror(false, true);
            currentImage.setImageType(OF_IMAGE_GRAYSCALE);
            currentImage.resize(currentImage.getWidth()/mod, currentImage.getHeight()/mod);

            unsigned char * pixels = resultImage.getPixels();

            int brightest = 0;

            for(int x = 0; x < currentImage.width; x++){
                    for(int y = 0; y < currentImage.height; y++){
                            int i = x + y * currentImage.width;
```

```
                              int color = blur(&currentImage, x, y, 5);
                              int prevColor = prevImage.getPixels()[i];

                              if((color + threshold <= prevColor || color - threshold >= prevColor)){
                                      pixels[i] = 255;
                              } else {
                                      pixels[i] = 0;
                              }
                      }
              }

              resultImage.update();
      }
}

//------------------------------------------------------------
void testApp::draw() {
    ofSetColor(255, 255, 255);
    resultImage.draw(640, 0,
                            320, 240);
    currentImage.draw(320, 0,
                            320, 240);
    prevImage.draw(0, 0,
                            320, 240);

}


//------------------------------------------------------------
void testApp::exit() {
    kinect.setCameraTiltAngle(0); // zero the tilt on exit
    kinect.close();
}


//------------------------------------------------------------
float testApp::blur(ofImage* img, int x, int y, int blurSize){
    float greyLevel = 0;

    unsigned char* pixels = img->getPixels();

    int numPixels = 0;

  for(int dx = -blurSize; dx <= blurSize; dx++){
                for(int dy = -blurSize; dy <= blurSize; dy++){

                      int newX = ofClamp((dx + x), 0, img->getWidth()  - 1);
                      int newY = ofClamp((dy + y), 0, img->getHeight() - 1);

                      numPixels++;
```

```
                        int i = (newX + newY * img->getWidth());

                        greyLevel += pixels[i];

                }
        }

        greyLevel = greyLevel/numPixels;

        return greyLevel;
}

//------------------------------------------------------------
void testApp::keyPressed (int key) {
        switch (key) {
                case OF_KEY_RETURN:
                        prevImage.setFromPixels(currentImage.getPixels(), currentImage.getWidth(),
currentImage.getHeight(), OF_IMAGE_GRAYSCALE, true);
                        break;
        }
}
```

Background subtraction is very effective in identifying whether a part of the image has changed since the initial background image was captured. Unfortunately, many factors can cause the scene captured by the camera to change over time.

Look at the example image shown in Figure 4-8.

Figure 4-8. *A background image (left), the same scene with lighting changes (center), the resulting image (right)*

Though there are no new objects in the scene, the background subtraction interprets the light changes as new entities. One of the benefits of using the Kinect's depth image is that the depth image uses infrared light, rather than visible light, to analyze the scene. Thus, changes in visible light do not affect how the background subtraction algorithm works, (see Figure 4-9).

Figure 4-9. A background depth image (left), the same depth image with a person in it and lighting changes (center), and the resulting image (right)

Again, by simply replacing the RGB image with the depth image, we resolve some of these issues.

1. Comment out the following line:

```
//greyImage.setFromPixels(kinect.getPixels(), kinect.getWidth(),
kinect.getHeight(), OF_IMAGE_COLOR, true);
```

2. Add in this one:

```
greyImage.setFromPixels(kinect.getDepthPixels(), kinect.getWidth(),
kinect.getHeight(), OF_IMAGE_GRAYSCALE, true);
```

This means that we process images not only when light conditions change but also in the dark!

Frame Differencing

Frame differencing shows the difference between the current frame's image and the previous frame's image and is often used to detect motion in a scene. When an object is located in one position in one frame, the color value of the pixels in those locations changes. In Figure 4-10 shows the frame difference between two frames in which the user's hand moves form a raised to a lowered position.

Figure 4-10. User with a hand raised (left), with a lowered hand (center), and the difference (right)

Thus, frame differencing can be used to detect motion in a scene. Listing 4-7 shows how to implement frame differencing with an RGB image:

Listing 4-7. *Frame Differencing Algorithm*

```
#include "testApp.h"
int mod = 4;
float threshold = 150;

//----------------------------------------------------------------
void testApp::setup() {
        kinect.init();
        kinect.setVerbose(true);
        kinect.open();

        prevImage.allocate(kinect.width/mod, kinect.height/mod, OF_IMAGE_GRAYSCALE);
        resultImage.allocate(kinect.width/mod, kinect.height/mod, OF_IMAGE_GRAYSCALE);
        ofSetFrameRate(60);

        // zero the tilt on startup
        angle = 0;
        kinect.setCameraTiltAngle(angle);
}

//----------------------------------------------------------------
void testApp::update() {
        ofBackground(100, 100, 100);
        threshold = ofMap(mouseX, 0, ofGetViewportWidth(), 0, 255, 255);
        kinect.update();
        if(kinect.isFrameNew())  // there is a new frame and we are connected
        {

                currentImage.setFromPixels(kinect.getPixels(), kinect.getWidth(),
kinect.getHeight(), OF_IMAGE_COLOR, true);
                //currentImage.setFromPixels(kinect.getDepthPixels(), kinect.getWidth(),
kinect.getHeight(), OF_IMAGE_GRAYSCALE, true);
                currentImage.mirror(false, true);
                currentImage.setImageType(OF_IMAGE_GRAYSCALE);
                currentImage.resize(currentImage.getWidth()/mod,
currentImage.getHeight()/mod);

                unsigned char * pixels = resultImage.getPixels();

                int brightest = 0;

                for(int x = 0; x < currentImage.width; x++){
                        for(int y = 0; y < currentImage.height; y++){
                                int i = x + y * currentImage.width;

                                int color = blur(&currentImage, x, y, 5);
                                int prevColor = prevImage.getPixels()[i];

                                if((color + threshold <= prevColor || color - threshold >=
prevColor)){
```

81

```
                                            pixels[i] = 255;
                             } else {
                                            pixels[i] = 0;
                             }
                      }
               }

               resultImage.update();
               prevImage.setFromPixels(currentImage.getPixels(), currentImage.getWidth(),
currentImage.getHeight(), OF_IMAGE_GRAYSCALE, true);
        }
}

//----------------------------------------------------------------
void testApp::draw() {
        ofSetColor(255, 255, 255);
        resultImage.draw(640, 0,
                                320, 240);
        currentImage.draw(320, 0,
                                320, 240);
        prevImage.draw(0, 0,
                                320, 240);
}


//----------------------------------------------------------------
void testApp::exit() {
        kinect.setCameraTiltAngle(0); // zero the tilt on exit
        kinect.close();
}


//----------------------------------------------------------------
float testApp::blur(ofImage* img, int x, int y, int blurSize){
        float greyLevel = 0;
        unsigned char* pixels = img->getPixels();
        int numPixels = 0;
    for(int dx = -blurSize; dx <= blurSize; dx++){
                for(int dy = -blurSize; dy <= blurSize; dy++){


                        int newX = ofClamp((dx + x), 0, img->getWidth()  - 1);
                        int newY = ofClamp((dy + y), 0, img->getHeight() - 1);
                        numPixels++;
                        int i =  (newX + newY * img->getWidth());
                        greyLevel += pixels[i];


                }
        }

    greyLevel = greyLevel/numPixels;
    return greyLevel;
```

```
}
//-------------------------------------------------------------------
void testApp::keyPressed (int key) {
        switch (key) {
                case OF_KEY_UP:
                        angle++;
                        if(angle>30) angle=30;
                        kinect.setCameraTiltAngle(angle);
                        break;

                case OF_KEY_DOWN:
                        angle--;
                        if(angle<-30) angle=-30;
                        kinect.setCameraTiltAngle(angle);
                        break;
                case OF_KEY_RETURN:
                        prevImage.setFromPixels(currentImage.getPixels(),
currentImage.getWidth(), currentImage.getHeight(), OF_IMAGE_GRAYSCALE, true);
                        break;
        }
}

//-------------------------------------------------------------------
void testApp::mouseMoved(int x, int y) {
}

//-------------------------------------------------------------------
void testApp::mouseDragged(int x, int y, int button)
{}

//-------------------------------------------------------------------
void testApp::mousePressed(int x, int y, int button)
{
        cout << threshold << endl;
}

//-------------------------------------------------------------------
void testApp::mouseReleased(int x, int y, int button)
{}

//-------------------------------------------------------------------
void testApp::windowResized(int w, int h)
{}
```

However, motion is not the only reason for a change between images. Similar to background subtraction, frame differencing can be confused by changes in lighting conditions. While frame differencing handles gradual changes in lighting conditions more successfully than background subtraction, it can still be fooled by a sudden change in lighting. Frame differencing will be unable to distinguish sudden light changes from motion. A flash from a camera, turning on or off a light, or a screen in the camera's field of view changing from one image to another will all cause a difference in frames resulting from color change unrelated to motion.

Frame differencing will also have issues distinguishing similar colors. If the area of change in the new image is a similar color to the unchanged area, frame differencing will not be able recognize motion in that area. Figure 4-11 shows an example of a frame differencing failure due to a too-close match in foreground and background colors.

Figure 4-11. Frame differencing failing when the user is wearing a white shirt in front of a white background

Fortunately, the Kinect's depth image will not suffer from these issues. Because it does not use visible light, changes in lighting conditions will not cause false motion in frame differencing. What's more, a frame where motion occurs with an object of a similar color to the background will not cause any issues.

Once more, we replace the RGB image with the depth image.

1. Comment out this line:

```
//greyImage.setFromPixels(kinect.getPixels(), kinect.getWidth(),
kinect.getHeight(), OF_IMAGE_COLOR, true);
```

2. Add this one:

```
greyImage.setFromPixels(kinect.getDepthPixels(), kinect.getWidth(),
kinect.getHeight(), OF_IMAGE_GRAYSCALE, true);
```

In Figure 4-12, you can see the results from the depth image do not suffer from the same issues as the results from the RGB image.

Figure 4-12. Image 1, image 2, and image 3 showing the frame difference using the depth camera

Combining Frame Differencing with Background Subtraction

One issue you may notice with frame differencing is that it finds differences in a moving object's position from one frame to the next. See Figure 4-13 for an example.

Figure 4-13. *Image of a double image with hand*

Technically, the double image in Figure 4-13 is correct, as motion occurred where the hand was in the previous frame (it left that area) and where it is in the current frame (it entered that area), but showing this double image is not desirable for all situations. To limit the detection of motion to an object's current location, we can combine background subtraction with frame differencing.

First, as shown in Listing 4-8, we use the same technique for background subtraction as above to create a black and white image of the new elements in the scene. Figure 4-14 shows the resulting image.

Listing 4-8. *Background Subtraction Combined with Frame Differencing*

```
    currentImage.setFromPixels(kinect.getDepthPixels(), kinect.getWidth(),
kinect.getHeight(), OF_IMAGE_GRAYSCALE, true);
    currentImage.mirror(false, true);
    currentImage.setImageType(OF_IMAGE_GRAYSCALE);
    currentImage.resize(currentImage.getWidth()/mod, currentImage.getHeight()/mod);

    unsigned char * pixels = resultBgSubImage.getPixels();

    for(int x = 0; x < currentImage.width; x++){
            for(int y = 0; y < currentImage.height; y++){
                    int i = x + y * currentImage.width;

                    int color = blur(&currentImage, x, y, 5);
                    int prevColor = bgImage.getPixels()[i];

                    if((color + threshold <= prevColor || color - threshold >=
prevColor)){
```

```
                                    pixels[i] = 255;
                        } else {
                                    pixels[i] = 0;
                        }
                }
        }


        resultBgSubImage.update();


        pixels = resultCombinedImage.getPixels();

        for(int x = 0; x < resultCombinedImage.width; x++){
                for(int y = 0; y < resultCombinedImage.height; y++){
                        int i = x + y * resultBgSubImage.width;

                        int color = resultBgSubImage.getPixels()[x + resultBgSubImage.width *
y];
                        int prevColor = prevBgSubImage.getPixels()[i];

                        if((color + threshold <= prevColor || color - threshold >=
prevColor)){
                                    pixels[i] = 255;
                        } else {
                                    pixels[i] = 0;
                        }
                }
        }


                resultCombinedImage.update();
```

Figure 4-14. Background subtraction combined with frame differencing image

Now, store this image for comparison in the future. Listing 4-9 shows one way in which to do that.

Listing 4-9. Code for Storing an Image as prevFrame

```
prevBgSubImage.setFromPixels(resultBgSubImage.getPixels(),
resultBgSubImage.getWidth(), resultBgSubImage.getHeight(), OF_IMAGE_GRAYSCALE,
true);
```

When the next frame comes in, we again apply the background subtraction algorithm to this frame. We then compare that new image with the previous background subtraction image. If a pixel changes from black to white between the previous frame and the current frame, we set it to white, (see Figure 4-15 for an example). Otherwise, we set it to black. This way, only the pixels that have a new object present will be perceived as having motion:

Figure 4-15. Image of combined techniquies

Summary

Several image processing approaches can yield improved results when using the Kinect's depth image. In some cases, using the depth image will provide completely different results from the standard image, which can be useful for tracking different interactions. This chapter presented just a few examples of image processing techniques that can be helpful with your Kinect hacks.

Libraries can also be useful in successfully executing many of these types of image processing techniques. However, it is always good to be able to understand how to do pixel-level image processing yourself. This will not only help you understand how the libraries work, but allow you to write your own versions of image processing that are specifically tailored to the interaction you are attempting to create.

CHAPTER 5

Gesture Recognition

Vision-based gesture recognition is one of the most powerful techniques available for human-computer interaction. In this chapter, we will explore how a combination of the image processing methods discussed in Chapter 4 can be used to turn any surface into a multitouch sensor. We will also look at how some widely used gestures such as pinch-to-zoom can be extracted from the data.

What Is a Gesture?

While this question may seem overly simple at first glance, it's not. Most people will probably think of a gesture as something like a wave of the hand, used in conversation. If you own a smartphone, you might consider pinch-to-zoom to be a gesture. Maybe even a DJ scratching over a record is performing a gesture?

These examples obviously suggest that there can't be one single answer to this question; the only definition broad enough would perhaps be "any motion performed by a human to convey some information." However, by this definition, even a mouse click is a gesture. Because detecting each and every nuance of human motion is beyond the scope of this chapter and also beyond the capabilities of the Kinect, we will focus on one scenario: detecting touch interaction on arbitrary surfaces.

Another option would be to use the entire body pose of the user as input; this is what the Kinect does in its original purpose as an Xbox controller device. However, detection and tracking of a simplified skeleton model of the user requires the use of the NITE software package, which we won't discuss in this book for the reasons outlined in Chapter 2.

Multitouch Detection

While the Kinect had the original purpose of tracking a free-standing person viewed from the front, it is quite easy to repurpose the depth data into something completely different, such as touch detection.

Consider a display surface, such as a projection screen or an LCD display, with a Kinect camera looking onto the surface from the same side as the user. Generally speaking, projection screens should not pose any problems in this scenario, while LCDs may cause some issues with image noise, particularly those with a glossy surface. When in doubt, just give it a try; most nonglossy LCD screens work surprisingly well. Mounting the Kinect slightly off-axis to avoid interfering with the user may be necessary, but this doesn't negatively influence our setup.

With no user present, the Kinect will now simply deliver an image of a flat surface, that is, one with approximately equal grey levels. Because of the design of the infrared emitter and detector, the screen content will have no noticeable effect on the depth data. We will call this image the background image.

Now, imagine a user in front of the display, touching the surface with her or his fingers. If the display is not occluded by the user's body (the reason for mounting the Kinect off-axis), the image will contain the flat display surface with the hand superimposed. As the hand is slightly closer to the camera than the surface, its pixel values will be lower by a small amount.

When we subtract the background image from the current depth camera image, the result will be an image that contains everything that is closer to the camera than the display, that is, the user's fingers and hand.

In the next step, we will now apply a modified threshold filter to the image. Usually, threshold filters set all pixels in an image with a value above the threshold to white, the rest to black. However, in this case, we need a slightly more complex filter. Since we only want to detect the fingertips that are touching the surface, we will have to filter out everything that is outside a small distance range, starting about 5 mm above the display (h_{min}) and ending at about 20 mm (h_{max}), as illustrated in Figure 5-1.

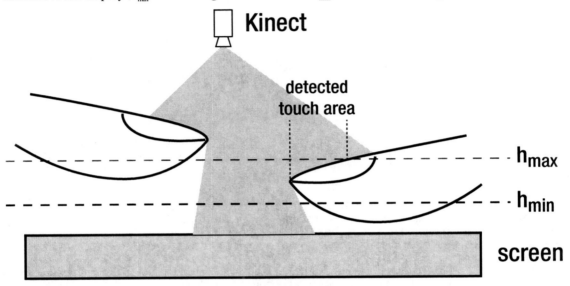

Figure 5-1. *Diagram of one fingertip touching the display surface and one hovering above.*

Applying that filter now results in a mostly black image with some white spots where fingers (or other objects with similar dimensions) touch the surface. Note that it will not be possible to detect the exact moment of touching, since we have to account for image noise and users with different finger thickness. Consequently, we will have to account for some margin of error in our threshold values.

Next, we will have to group these white spots together: from the point of view of the computer, they still consist of individual white pixels with no connection to their neighbors. This is achieved by so-called *connected component analysis*, a process that unifies coherent areas of identical color into components, or *blobs*.

The final step in this process is to track the fingertip locations. This means assigning a unique ID to each blob representing a touch point that stays with that point as long as the fingertip touches the surface. Having such IDs makes it possible to track the motion of each touch point over time, which is a crucial requirement for determining the gesture(s) executed by the user.

Summing up, the algorithm for multitouch detection using the Kinect consists of the following steps:

1. Acquire a camera image, and store it as the background image (see Figure 5-2 a).

2. While the program is running

   a. Acquire the camera image, and subtract background image (see Figure 5-2 b).

   b. Apply the threshold filter: a <= value <= b (see Figure 5-2 c).

   c. Identify connected components.

   d. Assign and track component IDs (see Figure 5-2 d).

   e. Calculate gesture values (e.g., zoom factor).

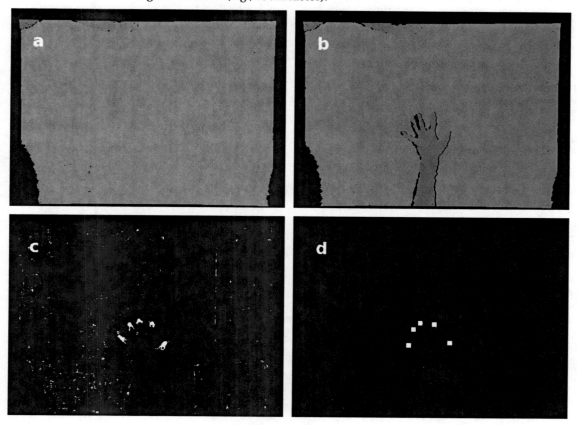

Figure 5-2. The four image processing steps: a) background image, b) current camera image, c) threshold difference image, d) detected components

We will now discuss each of those steps in turn.

Acquiring the Camera Image, Storing the Background, and Subtracting

These steps are conceptually quite simple and have already been explained extensively in the previous chapter. They are included in Listing 5-1.

Applying the Threshold Filter

In this step, we have to apply the modified threshold filter. Note that we are using fixed threshold values for both the lower and the upper threshold—depending on your setup and the distance between Kinect and display, these may have to be adjusted. Although OpenCV offers two different functions for applying threshold operators to images, neither of them is suitable for the dual threshold we are going to apply. Consequently, we will execute this operation directly on the image data as shown in Listing 5-1.

Listing 5-1. Background Acquisition and Double Threshold Calculation (Using C++)

```
// 16-bit greyscale image buffers
Mat rawdepth(Size(640,480),CV_16UC1);
Mat background(Size(640,480),CV_16UC1);
Mat difference(Size(640,480),CV_16UC1);

// 8-bit greyscale image buffer
Mat thresh(Size(640,480),CV_8UC1);

// control variables
bool die = false;
int reset_bg = 5;

// threshold values
unsigned char lower_val =  8;
unsigned char upper_val = 20;


// main loop
while (!die) {

        // update background to current camera image

        // current image is updated by libfreenect callback (not shown)
        if (reset_bg) {
                printf("resetting background...\n");
                rawdepth.copyTo( background );
                reset_bg--;
        }

        // subtract current image from background
        subtract( background, rawdepth, difference );

        // apply double threshold in simple loop over all pixels
        uint16_t* diffdata = (uint16_t*)difference.data;
```

```
uint8_t*  threshdata = thresh.data;

for (int i = 0 ; i < 640*480; i++) {
        if ((diffdata[i] >= lower_val) && (diffdata[i] <= upper_val))
                threshdata[i] = 255;
        else
                threshdata[i] = 0;
}

// display result
imshow("Chapter 5 Example",thresh);

// check for keypress
char k = cvWaitKey(5);

if ( k == 27 ) die = true;   // ESC key
if ( k == 32 ) reset_bg = 1; // space
}
```

This example includes the option to reset the background image by pressing the space bar, which might be necessary when your image changes significantly—for example, when the Kinect is moved with respect to the display. Also, the Kinect often needs a certain amount of startup time to stabilize the depth image. In this case, resetting the background image after a couple of seconds might also help.

Identifying Connected Components

In all likelihood, you will already have used a connected component algorithm without even realizing it. The most common application for this type of algorithm is the flood fill tool used in most image editors. This algorithm requires a start location, a target color, and a replacement color. In our case, the target color is white (we are looking for white blobs), and the replacement color is black (we only want to process every blob once, so we delete it by filling it with black). The algorithm is described in pseudocode in Listing 5-2.

Listing 5-2. Connected Components Algorithm (Pseudocode)

```
color old_color, new_color
connected_component( x, y ):
        if pixel(x,y) != old_color: return
        pixel(x,y) = new_color
        connected_component(x+1,y)
        connected_component(x - 1,y)
        connected_component(x,y+1)
        connected_component(x,y - 1)
```

Because of its recursive nature and the potential limitations, such as overrunning the stack, the algorithm is rarely implemented in this fashion. However, for understanding its purpose, this form is very well suited. Nevertheless, we will use OpenCV's implementation (cv::findContours) in Listing 5-3, since it is significantly faster than any native implementation.

After each connected component has been processed, we have a list of these components in the camera image. For each component, we can now determine the following:

- Location

- Area

- Approximate orientation and shape

This information is usually calculated from the *central moments* (mathematical properties of an area or shape) of each component. To this end, we can use the OpenCV function cv::moments on each of the component contours found in the previous step.

After we have retrieved location and size for every component contour, we can use this information to remove noise and other unwanted contours from the image. Since we know that we are looking for fingertips in the filtered image, we can give a rough estimate for the area of a fingertip, and depending on the distance between Kinect and interaction surface, we can assume an area range between roughly 25 and 150 pixels. Smaller components are probably noise, while larger components are likely flat objects (e.g., a mobile phone on the surface) that should not be identified as touch points.

Listing 5-3. *Finding Connected Components (Using C++)*

```cpp
// structure for describing a touch location
struct location {
        Vec2f position;
        Vec2f speed;
        Vec2f start_pos;
        bool tracked;
        int id,size;
};

// size thresholds
int min_size =  25;
int max_size = 150;

// storage for contours & touch locations
vector< vector<Point> > contours;
vector< location > touches;

// extract contours of connected components
findContours( thresh, contours, CV_RETR_LIST, CV_CHAIN_APPROX_SIMPLE );

// calculate moments and touch points
vector< vector<Point> >::iterator vec;
for (vec = contours.begin(); vec != contours.end(); vec++) {

        // create Moments object
        Mat points = Mat(*vec);
        Moments img_moments = moments( points );
        double area = img_moments.m00;

        // area of component within range?
        if ((area > min_size) && (area < max_size)) {
                location temp_loc;
                temp_loc.position=Vec2f( img_moments.m10/area,img_moments.m01/area );
                temp_loc.size = area;
```

```
            touches.push_back( temp_loc );
        }
}
```

Assigning and Tracking Component IDs

The next, very important step is to track the individual touch locations as the fingertips move over the surface. Without this step, we would have a list of touch locations for each camera image but no way of knowing whether the first touch location in image i actually belongs to the same finger as the first touch location in image i+1. However, to determine the gestures executed by the user, we will need the full trajectories of the touch locations over multiple frames. Consequently, we will now have to correlate the touch locations between the current image and the last one.

To avoid cluttering Listing 5-4 with too many details, we will continue where the previous code example ended. We use a simple object type location (already introduced previously) to describe every touch location. This object contains current touch position, speed relative to the previous frame, initial location, component size and a unique numeric identifier. Also, we will use a second list of locations prev_locations containing the data from the previous frame.

Listing 5-4. *Touch Location Tracking (Using C++)*

```cpp
// counter for new IDs
int next_id = 1;

// touch locations from previous frame
vector< location > prev_touches;

// assign new IDs to all new touch locations
for (vector<location>::iterator loc = touches.begin(); loc != touches.end(); loc++) {
        loc->start_pos = loc->position;
        loc->speed = Vec2f(0,0);
        loc->tracked = false;
        loc->id = next_id;
        next_id++;
}

// see if we can match any old locations with the new ones
for (vector<location>::iterator prev_loc = prev_touches.begin(); prev_loc !=
prev_touches.end(); prev_loc++) {

        // predict new position
        Vec2f predict = prev_loc->position + prev_loc->speed;
        double mindist = norm(prev_loc->speed) + 5; // minimum search radius: 5 px

        location* nearest = NULL;

        // search closest new touch location (that has not yet been assigned)
        for (vector<location>::iterator loc = touches.begin(); loc != touches.end(); loc++) {

                if (loc->tracked) continue;
                Vec2f delta = loc->position - predict;
                double dist = norm(delta);
```

```
                    if (dist < mindist) {
                            mindist = dist;
                            nearest = &(*loc);
                    }
            }

            // assign data from previous location
            if (nearest != NULL) {
                    nearest->id = prev_loc->id;
                    nearest->speed = nearest->position - prev_loc->position;
                    nearest->start_pos = prev_loc->start_pos;
                    nearest->tracked = true;
            }
    }

    // paint touch locations into image
    for (vector<location>::iterator loc = touches.begin(); loc != touches.end(); loc++) {
            char textbuffer[32]; snprintf(textbuffer,32,"%d",loc->id);
            Point v1 = Point(loc->position) - Point(5,5);
            Point v2 = Point(loc->position) + Point(5,5);
            rectangle( thresh, v1, v2, 255, CV_FILLED );
            putText( thresh, textbuffer, v2, FONT_HERSHEY_SIMPLEX, 1, 255 );
    }
}
```

This algorithm first assigns increasing IDs to all new locations. In the main loop, the predicted new location for every previous location is calculated from the previous position and speed. In a radius proportional to the speed of the touch location, the closest new location is found. This location is assigned the same ID as the previous location, and the new location is removed from further consideration. After the algorithm has completed, all touch locations near the predicted new position of a location from the previous frame have been assigned the corresponding IDs, and all new touch locations have been assigned new, previously unused IDs.

Calculating Gestures

The final step in our program is now to determine the gestures actually executed by the user. For this example, we will focus on three gestures that are widely used on multitouch-capable devices and that are, consequently, well known to most tech-savvy people:

- Tapping and dragging to *move* objects

- Rotating two or more fingers with respect to each other to *rotate* objects

- Pinch-to-zoom to *scale* objects

It is important to note in this context that several gestures can happen concurrently. For example, the user might drag fingers apart horizontally and move them vertically at the same time, resulting in move and zoom actions occurring simultaneously.

For our gesture detector, we will use all detected touch locations. Of course, this means that several people interacting together with an image (e.g., a map) might interfere with each other. The best way to avoid such problems depends heavily on the specific application and might be to only use the first two touch locations or to group touches into separate screen areas.

Detecting Motion

The easiest gesture to calculate is the gesture to move objects. Here, we rely on the start_loc field introduced earlier, which describes the initial position of the touch location. For every touch location, we calculate the vector between the initial and current positions. Then, we simply calculate the average motion vector by summing these up and dividing the result vector by the number of touch locations (see Listing 5-5).

Listing 5-5. Determining Motion (Using C++)

```
double touchcount = touches.size();

// determine overall motion
Vec2f motion(0,0);

for (vector<location>::iterator loc = touches.begin(); loc != touches.end(); loc++) {
        Vec2f delta = loc->position - loc->start_pos;
        motion = motion + delta;
}

// avoid division by zero - otherwise, calculate average
if (touchcount > 0) {
        motion[0] = motion[0] / touchcount;
        motion[1] = motion[1] / touchcount;
}
```

The result is a vector representing the average motion of all touch locations.

Rotation

Next, we will focus on the gesture for rotation. This gesture is very intuitive, since it consists of a similar motion to what most people would use to rotate a sheet of paper lying on a table: put down two or more fingers on the object, and rotate them relative to each other.

We start by calculating the initial and current centroid of all touch locations; all further calculations will be relative to these two centroids. Next, we determine the angle by which each touch location has moved around the centroid relative to its start position and finally calculate the average angle. The code is in Listing 5-6.

Listing 5-6. Determining Rotation (Using C++)

```
// determine initial/current centroid of touch points (for rotation & scale)
Vec2f centroid_start(0,0);
Vec2f centroid_current(0,0);

for (vector<location>::iterator loc = touches.begin(); loc != touches.end(); loc++) {
        centroid_start = centroid_start + loc->start_pos;
        centroid_current = centroid_current + loc->start_pos;
}

if (touchcount > 0) {
        centroid_start[0] = centroid_start[0] / touchcount;
        centroid_start[1] = centroid_start[1] / touchcount;
        centroid_current[0] = centroid_current[0] / touchcount;
        centroid_current[1] = centroid_current[1] / touchcount;
}

// calculate rotation
double angle = 0.0;

for (vector<location>::iterator loc = touches.begin(); loc != touches.end(); loc++) {

        // shift vectors to centroid-based coordinates
        Vec2f p0 = loc->start_pos - centroid_start;
        Vec2f p1 = loc->position  - centroid_current;

        // normalize vectors
        double np0 = norm(p0), np1 = norm(p1);
        p0[0] = p0[0]/np0; p0[1] = p0[1]/np0;
        p1[0] = p1[0]/np1; p1[1] = p1[1]/np1;

        // scalar product: determine rotation angle
        double current_angle = acos( p0.ddot(p1) );

         // cross product: determine rotation direction
        Mat m0 = ( Mat_<double>(3,1) << p0[0], p0[1], 0 );
        Mat m1 = ( Mat_<double>(3,1) << p1[0], p1[1], 0 );

        Mat cross = m0.cross(m1);

        if (cross.at<double>(2,0) < 0)
                current_angle = -current_angle;

        angle += current_angle;
}

if (touchcount > 0)
        angle = angle / touchcount;
```

The result is the angle in radians by which the touch locations have rotated on average. Note that the result only makes sense when at least two touch locations are present.

Scale

The last gesture we will look at is often called pinch-to-zoom, and the code for it is in Listing 5-7. Thanks to the iPhone, this is probably the most widely used multitouch gesture, although it may not appear to be as intuitive as the rotation gesture. There are a lot of similarities between these two gestures, and they can also be easily used in parallel.

Again, we start by calculating the initial and current centroid of all touch locations. In contrast to the previous gesture, we now determine the change in distance relative to the centroid as a floating-point number. Numbers below 1.0 indicate shrinking (fingers are moving closer together), and numbers above 1.0 indicate enlarging (fingers are moving apart). We will reuse the centroid_start and centroid_current variables from the previous example.

Listing 5-7. *Determining Scaling (Using C++)*

```
double scale = 0.0;

for (vector<location>::iterator loc = touches.begin(); loc != touches.end(); loc++) {

        // shift vectors to centroid-based coordinates
        Vec2f p0 = loc->start_pos - centroid_start;
        Vec2f p1 = loc->position  - centroid_current;

        double relative_distance = norm(p0) / norm(p1);
        scale += relative_distance;
}

if (touchcount > 0)
        scale = scale / touchcount;
```

The result is a floating-point number indicating the relative size change of the touch locations. This result is also only useful when at least two touch locations are present.

When directly using the result values from these three gestures to control an image (e.g., a map), some jitter will often be noticeable. This is caused by noise in the image, which, in turn, causes the touch locations to move slightly between frames, even if the user is keeping perfectly still. An easy fix for this type of problem is to introduce small thresholds below which the gesture result is not used (e.g., only rotate the map when the angle is above 3°).

However, whether this approach is really an improvement again strongly depends on your application and your users—experiment!

Creating a *Minority Report*—Style Interface

In addition to the iPhone, Hollywood has also played an important part in making gestural interfaces known to a wider audience. In particular, the movie *Minority Report* (based on a short story by Philip K. Dick) has done a lot to make these interfaces cool by showing Tom Cruise zooming, browsing, and scrolling through a stack of videos using expansive hand gestures. (What the movie doesn't show is that even a fit, trained actor like Tom Cruise wasn't able to use this kind of interface longer than 10 minutes without needing a break. This feeling of tiredness is sometimes called the *gorilla arm syndrome*.)

Fitness issues notwithstanding, we'll look at how the software developed in this chapter can be easily modified to create a similar interface to the one used in the movie. Instead of having the Kinect looking down on a display surface, consider a setup much like the one the Kinect was originally intended for: standing on top of a large screen, looking toward the user. In the previous chapter, you saw how the brightest pixel in the depth image corresponds to the closest point in the Kinect's field of view. When we look at a user standing in front of the camera with stretched-out hands pointing toward the screen, the closest point will be part of one of the hands, as illustrated in Figure 5-3.

Now, all we have to do is take the depth value of this point, add a margin corresponding to approximately 10 cm, and use this value as threshold value in our algorithm from above; all the other steps can be used as before. You will probably also have to adjust the size threshold for the connected components, because we are no longer looking for fingertips but for entire hands, which are significantly bigger. However, when the user lowers her or his hands, the closest point will be somewhere on the body, leading to a single, very large connected component. Consequently, a size filter between 500 and 5000 pixels might be appropriate in this case. Finally, it is also possible to drop the background subtraction step entirely, because the threshold value will be generated on the fly.

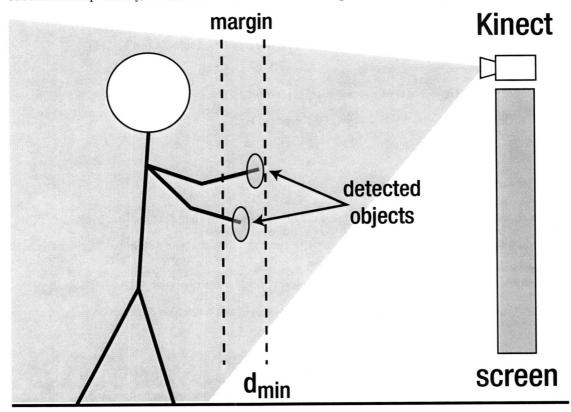

Figure 5-3. A user in front of the screen, using a Minority Report*–style interface*

Considering Shape Gestures

Sometimes when gestures are mentioned, people think more about shape gestures than about the multitouch gestures used in this chapter. The term "shape gestures" describes shapes, such as letters, that are traced with the finger or a pen. However, shape gestures do not contradict multitouch gestures: both types of gestures can coexist easily. A very simple approach to doing so would be to require two or more fingers for all motion gestures (rotate and scale already need at least two touch points in any case) and try to detect shape gestures when only one finger is used.

A widely used method for detecting these gestures is the *$1 Gesture Recognizer*, which was introduced in an academic paper by J. Wobbrock, A. Wilson, and Y. Li.

■ **Note** Gestures without libraries, toolkits or training: a $1 recognizer for user interface prototypes (ACM, 2007).

Describing this algorithm is beyond the scope of this chapter, but numerous examples are available on the Web.

Summary

Using the Kinect's depth image and some of the image processing techniques outlined in the previous chapter, you can easily turn arbitrary surfaces into multitouch interfaces. This will work with regular displays such as LCD screens, but it is also possible to use projection screens, even arbitrarily shaped ones—consider a projected globe with touch interaction!

To this end, we've covered a number of basic image processing techniques which extract the coordinates of touch points on an arbitrary surface from the depth image. However, for a full gestural interface, extra work is required beyond retrieving the touch locations: it is also necessary to track them as the user's hand moves over the surface. Additional interpretation is required to extract meaning in the form of gestures from this motion data and use it to control an interface.

CHAPTER 6

Voxelization

In this chapter, we'll discuss voxelization and develop a people tracking system. We'll cover the following:

- What a voxel is

- Why you would want to voxelize your data

- How to make, manipulate, and use voxels

- Why people are rectangles

What Is a Voxel?

A *voxel* is the three-dimensional equivalent of a pixel—a box, rather than a point, in space that has a volume. Imagine taking an object and then decomposing into cubes, all of the same size. Or, if you'd prefer, building an object out of LEGOs or in Minecraft, much like a cubist painting. See Figure 6-1 for an example of a voxelized scene.

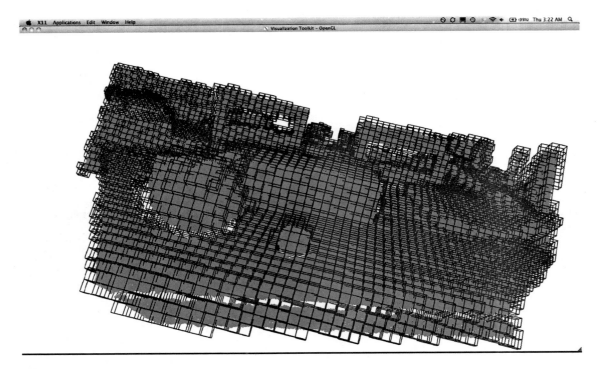

Figure 6-1. A voxelized scene

Why Voxelize Data?

There are two different reasons to voxelize a dataset. First, it reduces the size of the dataset, up to 75%—essential if you're going to be doing a large amount of work on your data in real time. This size reduction can create an outsized response in performance, given the computational complexity of some of the algorithms involved. Instead of simply destroying data, as if you decimated it, voxelization creates a smoothly moving average over the space, something that maintains as much data as possible while simultaneously lowering the state space size. For example, given a scene that's 3 meters by 3 meters by 3 meters and has 500,000 points evenly distributed, you can realize over a 50% reduction in points (216,000 vs. 500,000) with a 5-centimeter voxel size.

Second, voxelizing various datasets make them very easy to combine. This may not seem like a large benefit, but consider the multisensor fusion problem. Instead of relying on complex algorithms to handle fusing multiple point clouds, you simply need to address each potential voxel space of interest and calculate the probability that it is occupied given multiple sensors.

Voxels make your datasets easier to work with, but what's the next step? There are many possibilities for voxelized data, but we're going to cover background subtraction, people tracking, and collision detection. Background subtraction is cutting out the background of a scene (prerecorded) and isolating the new data. People tracking is using that new data (foreground) to track people and people shaped groups of voxels.

Voxelizing Data

Voxelizing in PCL is simply accomplished: the voxelizer is a filter, and it takes a point cloud as an input. It then outputs the voxelized point cloud, calculated for the centroid of the set of points voxelized. See Listing 6-1 for an example.

Listing 6-1. *Voxelizing Data*

```
#include <pcl/filters/voxel_grid.h>

//Voxelization
pcl::VoxelGrid<pcl::PointXYZRGB> vox;
vox.setInputCloud (point_cloud_ptr);
vox.setLeafSize (5.0f, 5.0f, 5.0f);
vox.filter (*cloud_filtered);
std::cerr << "PointCloud before filtering: " << point_cloud_ptr->width *
point_cloud_ptr->height << " data points (" << pcl::getFieldsList (*point_cloud_ptr) << ").";
std::cerr << "PointCloud after filtering: " << cloud_filtered->width *
cloud_filtered->height << " data points (" << pcl::getFieldsList (*cloud_filtered) << ").";
```

Let's break down the preceding listing:

```
pcl::VoxelGrid<pcl::PointXYZRGB> vox;
vox.setInputCloud (point_cloud_ptr);
```

The preceding two commands declare the filter (VoxelGrid) and set the input to the filter to be point_cloud_ptr.

```
vox.setLeafSize (5.0f, 5.0f, 5.0f);
```

The preceding command is the key part of the declaration: it sets the size of the window over which voxels are calculated or the voxel size itself. Given that our measurements from the Kinect are in millimeters, this corresponds to a voxel that is 5mm per side.

While this calculation seems quite easy, there's a problem: voxelizing in this manner disorders a point cloud. Instead of a set of points that's easy to access and manipulate set of points that's 640 high by 480 wide, you have a single line that's 307,200 points long. This makes accessing a particular point or even getting the proper set of indices for a group of points difficult.

A better method is to use *octrees*. Octrees divide space into recursive sets of eight octants and organize it as a tree, as shown in Figure 6-2. Only the leaves hold data, and you can easily access them. Also, there are several very fast functions for calculating useful metrics over an octree; for example, nearest neighbor searches are lightning fast. See Figures 6-3 and Figure 6-4 for two examples of octree divided objects.

Again, PCL provides functions to make this work simple; the octree module handles everything. Listing 6-2 will do the same work as Listing 6-1. It voxelizes the incoming point cloud about the centroids of points, dividing the world up neatly into 50-mm–per–side squares.

Listing 6-2. *Octree Creation*

```
#include <pcl/octree/octree.h>

float resolution = 50.0;
pcl::octree::OctreePointCloudVoxelCentroid<pcl::PointXYZRGB> octree(resolution);
// Add points from cloud to octree
octree.setInputCloud (pointcloud_ptr);
octree.addPointsFromInputCloud ();
```

The preceding code is very straightforward. Resolution defines the voxel size, in our case, 50 mm. We create the octree with standard octree and leaf types and set the points inside to be XYZRGB. We then set the input cloud to the octree as our cloud and call addPointsFromInputCloud to push them into the structure.

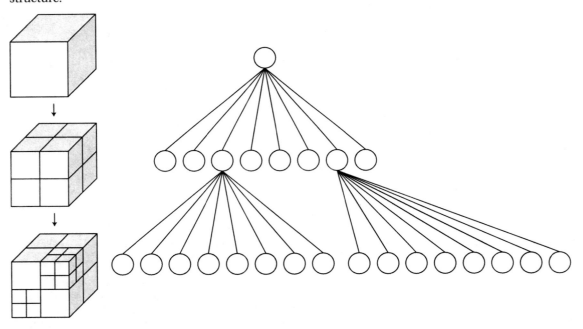

Figure 6-2. *How octrees are organized*

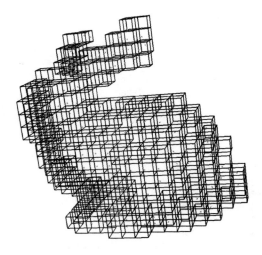

Figure 6-3. Octree bunny

Manipulating Voxels

In the end, we're really only interested in the leaf nodes of the octree that contain the data we're looking for. Normally, to reach all of the leaves, we'd have to traverse the entire tree end to end. However, PCL handily provides an iterator for leaf nodes, LeafNodeIterator:

```
// leaf node iterator
OctreePointCloud<PointXYZRGB>::LeafNodeIterator itLeafs (octree);
```

Let's work through a simple example. We're going to take an input cloud and display it voxelized with boxes drawn where there are filled voxels in space, as shown in Listing 6-3.

Listing 6-3. Drawing Voxel Boxes

```
viewer->removeAllShapes();
octreepc.setInputCloud(fgcloud);
octreepc.addPointsFromInputCloud();
//Get the voxel centers
double voxelSideLen;
std::vector<pcl::PointXYZRGB, Eigen::aligned_allocator<pcl::PointXYZRGB> > voxelCenters;
octreepc.getOccupiedVoxelCenters (voxelCenters);
std::cout << "Voxel Centers: " << voxelCenters.size() << std::endl;
voxelSideLen = sqrt (octreepc.getVoxelSquaredSideLen ());
//delete the octree
octreepc.deleteTree();
pcl::ModelCoefficients cube_coeff;
cube_coeff.values.resize(10);
cube_coeff.values[3] = 0; //Rx
cube_coeff.values[4] = 0; //Ry
cube_coeff.values[5] = 0; //Rz
cube_coeff.values[6] = 1; //Rw
cube_coeff.values[7] = voxelSideLen; //Width
cube_coeff.values[8] = voxelSideLen; //Height
cube_coeff.values[9] = voxelSideLen; //Depths
//Iterate through each center, drawing a cube at each location.
for (size_t i = 0; i < voxelCenters.size(); i++) {
    //Drawn from center
    cube_coeff.values[0] = voxelCenters[i].x;
    cube_coeff.values[1] = voxelCenters[i].y;
    cube_coeff.values[2] = voxelCenters[i].z;
    viewer->addCube(cube_coeff, boost::lexical_cast<std::string>(i));
}
```

Given these capabilities, you may ask, "What's the purpose of drawing voxel boxes?" While our ultimate goal is voxel-on-voxel collisions, our first goal is to subtract the background of our scene from our new data, as shown in Listing 6-4. First comes the code in block, and then we'll break down each step.

Listing 6-4. Background Subtraction

```
/*
 * This file is part of the OpenKinect Project. http://www.openkinect.org
 *
 * Copyright (c) 2010 individual OpenKinect contributors. See the CONTRIB file
 * for details.
 *
 * This code is licensed to you under the terms of the Apache License, version
 * 2.0, or, at your option, the terms of the GNU General Public License,
 * version 2.0. See the APACHE20 and GPL2 files for the text of the licenses,
 * or the following URLs:
 * http://www.apache.org/licenses/LICENSE-2.0
 * http://www.gnu.org/licenses/gpl-2.0.txt
 *
```

```
#include <iostream>
#include <libfreenect.hpp>
#include <pthread.h>
#include <stdio.h>
#include <string.h>
#include <cmath>
#include <vector>
#include <ctime>
#include <boost/thread/thread.hpp>
#include "pcl/common/common_headers.h"
#include "pcl/features/normal_3d.h"
#include "pcl/io/pcd_io.h"
#include "pcl/visualization/pcl_visualizer.h"
#include "pcl/console/parse.h"
#include "pcl/point_types.h"
#include <pcl/kdtree/kdtree_flann.h>
#include <pcl/surface/mls.h>
#include "boost/lexical_cast.hpp"
#include "pcl/filters/voxel_grid.h"
#include "pcl/octree/octree.h"

///Mutex Class
class Mutex {
public:
        Mutex() {
                pthread_mutex_init( &m_mutex, NULL );
        }
        void lock() {
                pthread_mutex_lock( &m_mutex );
        }
        void unlock() {
                pthread_mutex_unlock( &m_mutex );
        }

        class ScopedLock
        {
                Mutex & _mutex;
        public:
                ScopedLock(Mutex & mutex)
```

```
                                : _mutex(mutex)
                        {
                        _mutex.lock();
                        }
                        ~ScopedLock()
                        {
                                _mutex.unlock();
                        }
                };
        private:
                pthread_mutex_t m_mutex;
        };


        ///Kinect Hardware Connection Class
        /* thanks to Yoda---- from IRC */
        class MyFreenectDevice : public Freenect::FreenectDevice {
        public:
                MyFreenectDevice(freenect_context *_ctx, int _index)
                        : Freenect::FreenectDevice(_ctx, _index),
        depth(freenect_find_depth_mode(FREENECT_RESOLUTION_MEDIUM, FREENECT_DEPTH_REGISTERED).bytes),
        m_buffer_video(freenect_find_video_mode(FREENECT_RESOLUTION_MEDIUM,
        FREENECT_VIDEO_RGB).bytes), m_new_rgb_frame(false), m_new_depth_frame(false)
                        {

                        }
                //~MyFreenectDevice(){}
                // Do not call directly even in child
                void VideoCallback(void* _rgb, uint32_t timestamp) {
                        Mutex::ScopedLock lock(m_rgb_mutex);
                        uint8_t* rgb = static_cast<uint8_t*>(_rgb);
                        std::copy(rgb, rgb+getVideoBufferSize(), m_buffer_video.begin());
                        m_new_rgb_frame = true;
                };
                // Do not call directly even in child
                void DepthCallback(void* _depth, uint32_t timestamp) {
                        Mutex::ScopedLock lock(m_depth_mutex);
                        depth.clear();
                        uint16_t* call_depth = static_cast<uint16_t*>(_depth);
                        for (size_t i = 0; i < 640*480 ; i++) {
                                depth.push_back(call_depth[i]);
                        }
                        m_new_depth_frame = true;
                }
                bool getRGB(std::vector<uint8_t> &buffer) {
                        Mutex::ScopedLock lock(m_rgb_mutex);
                        if (!m_new_rgb_frame)
                        return false;
                        buffer.swap(m_buffer_video);
                        m_new_rgb_frame = false;
                        return true;
                }
```

```
        bool getDepth(std::vector<uint16_t> &buffer) {
                Mutex::ScopedLock lock(m_depth_mutex);
                if (!m_new_depth_frame)
                        return false;
                buffer.swap(depth);
                m_new_depth_frame = false;
                return true;
        }
private:
        std::vector<uint16_t> depth;
        std::vector<uint8_t> m_buffer_video;
        Mutex m_rgb_mutex;
        Mutex m_depth_mutex;
        bool m_new_rgb_frame;
        bool m_new_depth_frame;
};


///Start the PCL/OK Bridging

//OK
Freenect::Freenect freenect;
MyFreenectDevice* device;
freenect_video_format requested_format(FREENECT_VIDEO_RGB);
double freenect_angle(0);
int got_frames(0),window(0);
int g_argc;
char **g_argv;
int user_data = 0;


//PCL
pcl::PointCloud<pcl::PointXYZRGB>::Ptr cloud (new pcl::PointCloud<pcl::PointXYZRGB>);
pcl::PointCloud<pcl::PointXYZRGB>::Ptr bgcloud (new pcl::PointCloud<pcl::PointXYZRGB>);
pcl::PointCloud<pcl::PointXYZRGB>::Ptr voxcloud (new pcl::PointCloud<pcl::PointXYZRGB>);
pcl::PointCloud<pcl::PointXYZRGB>::Ptr fgcloud (new pcl::PointCloud<pcl::PointXYZRGB>);
float resolution = 50.0; //5 CM voxels
// Instantiate octree-based point cloud change detection class
pcl::octree::OctreePointCloudChangeDetector<pcl::PointXYZRGB> octree (resolution);
//Instantiate octree point cloud
pcl::octree::OctreePointCloud<pcl::PointXYZRGB> octreepc(resolution);

bool BackgroundSub = false;
bool hasBackground = false;
bool grabBackground = false;
int bgFramesGrabbed = 0;
const int NUMBGFRAMES = 10;
bool Voxelize = false;
bool Oct = false;
bool Person = false;
```

111

```cpp
unsigned int cloud_id = 0;

///Keyboard Event Tracking
void keyboardEventOccurred (const pcl::visualization::KeyboardEvent &event,
                           void* viewer_void)
{
  boost::shared_ptr<pcl::visualization::PCLVisualizer> viewer =
*static_cast<boost::shared_ptr<pcl::visualization::PCLVisualizer> *> (viewer_void);
  if (event.getKeySym () == "c" && event.keyDown ())
  {
    std::cout << "c was pressed => capturing a pointcloud" << std::endl;
    std::string filename = "KinectCap";
    filename.append(boost::lexical_cast<std::string>(cloud_id));
    filename.append(".pcd");
    pcl::io::savePCDFileASCII (filename, *cloud);
    cloud_id++;
  }

  if (event.getKeySym () == "b" && event.keyDown ())
  {
      std::cout << "b was pressed" << std::endl;
      if (BackgroundSub == false)
      {
          //Start background subtraction
          if (hasBackground == false)
          {
              //Grabbing Background
              std::cout << "Starting to grab backgrounds!" << std::endl;
              grabBackground = true;
          }
          else
              BackgroundSub = true;
      }
      else
      {
          //Stop Background Subtraction
          BackgroundSub = false;
      }
  }

  if (event.getKeySym () == "v" && event.keyDown ())
  {
      std::cout << "v was pressed" << std::endl;
      Voxelize = !Voxelize;
  }

  if (event.getKeySym () == "o" && event.keyDown ())
  {
      std::cout << "o was pressed" << std::endl;
      Oct = !Oct;
      viewer->removeAllShapes();
  }
```

```
  if (event.getKeySym () == "p" && event.keyDown ())
  {
      std::cout << "p was pressed" << std::endl;
      Person = !Person;
      viewer->removeAllShapes();
  }
}

// --------------
// -----Main-----
// --------------
int main (int argc, char** argv)
{
        //More Kinect Setup
        static std::vector<uint16_t> mdepth(640*480);
        static std::vector<uint8_t> mrgb(640*480*4);

        // Fill in the cloud data
        cloud->width    = 640;
        cloud->height   = 480;
        cloud->is_dense = false;
        cloud->points.resize (cloud->width * cloud->height);

        // Create and setup the viewer
        boost::shared_ptr<pcl::visualization::PCLVisualizer> viewer (new
pcl::visualization::PCLVisualizer ("3D Viewer"));
        viewer->registerKeyboardCallback (keyboardEventOccurred, (void*)&viewer);
        viewer->setBackgroundColor (0, 0, 0);
        viewer->addPointCloud<pcl::PointXYZRGB> (cloud, "Kinect Cloud");
        viewer->setPointCloudRenderingProperties
(pcl::visualization::PCL_VISUALIZER_POINT_SIZE, 1, "Kinect Cloud");
        viewer->addCoordinateSystem (1.0);
        viewer->initCameraParameters ();

        //Voxelizer Setup
        pcl::VoxelGrid<pcl::PointXYZRGB> vox;
        vox.setLeafSize (resolution,resolution,resolution);
        vox.setSaveLeafLayout(true);
        vox.setDownsampleAllData(true);
        pcl::VoxelGrid<pcl::PointXYZRGB> removeDup;
        removeDup.setLeafSize (1.0f,1.0f,1.0f);
        removeDup.setDownsampleAllData(true);

        //Background Setup
        pcl::RadiusOutlierRemoval<pcl::PointXYZRGB> ror;
        ror.setRadiusSearch(resolution);
        ror.setMinNeighborsInRadius(200);

        device = &freenect.createDevice<MyFreenectDevice>(0);
        device->startVideo();
        device->startDepth();
        boost::this_thread::sleep (boost::posix_time::seconds (1));
```

```
//Grab until clean returns
int DepthCount = 0;
while (DepthCount == 0) {
        device->updateState();
        device->getDepth(mdepth);
        device->getRGB(mrgb);
        for (size_t i = 0;i < 480*640;i++)
        DepthCount+=mdepth[i];
}

//--------------------
// -----Main loop-----
//--------------------
double x = NULL;
double y = NULL;
int iRealDepth = 0;
while (!viewer->wasStopped ())
{
        device->updateState();
        device->getDepth(mdepth);
device->getRGB(mrgb);

size_t i = 0;
size_t cinput = 0;
for (size_t v=0 ; v<480 ; v++)
{
    for ( size_t u=0 ; u<640 ; u++, i++)
    {
        iRealDepth = mdepth[i];
        freenect_camera_to_world(device->getDevice(), u, v, iRealDepth, &x, &y);
        cloud->points[i].x  = x;//1000.0;
        cloud->points[i].y  = y;//1000.0;
        cloud->points[i].z = iRealDepth;//1000.0;
        cloud->points[i].r = 255;//mrgb[i*3];
        cloud->points[i].g = 255;//mrgb[(i*3)+1];
        cloud->points[i].b = 255;//mrgb[(i*3)+2];
    }
}

std::vector<int> indexOut;
pcl::removeNaNFromPointCloud (*cloud, *cloud, indexOut);
if (grabBackground) {
    if (bgFramesGrabbed == 0)
    {
        std::cout << "First Grab!" << std::endl;
        *bgcloud = *cloud;
    }
    else
    {
        //concat the clouds
        *bgcloud+=*cloud;
    }
```

```
                bgFramesGrabbed++;
                if (bgFramesGrabbed == NUMBGFRAMES) {
                    grabBackground = false;
                    hasBackground = true;
                    std::cout << "Done grabbing Backgrounds - hit b again to subtract BG." <<
std::endl;
                    removeDup.setInputCloud (bgcloud);
                    removeDup.filter (*bgcloud);
                }
                else
                    std::cout << "Grabbed Background " << bgFramesGrabbed << std::endl;
                viewer->updatePointCloud (bgcloud, "Kinect Cloud");
        }
        else if (BackgroundSub) {
            octree.deleteCurrentBuffer();
            fgcloud->clear();

            // Add points from background to octree
            octree.setInputCloud (bgcloud);
            octree.addPointsFromInputCloud ();

            // Switch octree buffers
            octree.switchBuffers ();

            // Add points from the mixed data to octree
            octree.setInputCloud (cloud);
            octree.addPointsFromInputCloud ();

            std::vector<int> newPointIdxVector;

            //Get vector of point indices from octree voxels
            //which did not exist in previous buffer
            octree.getPointIndicesFromNewVoxels (newPointIdxVector, 1);

            for (size_t i = 0; i < newPointIdxVector.size(); ++i) {
                fgcloud->push_back(cloud->points[newPointIdxVector[i]]);
            }

            //Filter the fgcloud down
            ror.setInputCloud(fgcloud);
            ror.filter(*fgcloud);

            viewer->updatePointCloud (fgcloud, "Kinect Cloud");
        }
        else if (Voxelize) {
            vox.setInputCloud (cloud);
            vox.setLeafSize (5.0f, 5.0f, 5.0f);
            vox.filter (*voxcloud);
            viewer->updatePointCloud (voxcloud, "Kinect Cloud");
        }
        else
            viewer->updatePointCloud (cloud, "Kinect Cloud");
```

```
        viewer->spinOnce ();
    }
    device->stopVideo();
    device->stopDepth();
    return 0;
}
```

As you may notice, the first quarter of the code is just like the bridge code we built in Chapter 3. The background subtraction code is all new, and we'll be focusing on the functional part directly, in Listing 6-5.

Listing 6-5. Background Subtraction Function

```
if (grabBackground) {
    if (bgFramesGrabbed == 0)
    {
        std::cout << "First Grab!" << std::endl;
        *bgcloud = *cloud;
    }
    else
    {
        //concat the clouds
        *bgcloud+=*cloud;
    }
    bgFramesGrabbed++;
    if (bgFramesGrabbed == NUMBGFRAMES) {
        grabBackground = false;
        hasBackground = true;
        std::cout << "Done grabbing Backgrounds - hit b again to subtract BG." <<
std::endl;
        removeDup.setInputCloud (bgcloud);
        removeDup.filter (*bgcloud);
    }
    else
        std::cout << "Grabbed Background " << bgFramesGrabbed << std::endl;
    viewer->updatePointCloud (bgcloud, "Kinect Cloud");
}
else if (BackgroundSub) {
    octree.deleteCurrentBuffer();
    fgcloud->clear();

    // Add points from background to octree
    octree.setInputCloud (bgcloud);
    octree.addPointsFromInputCloud ();

    // Switch octree buffers
    octree.switchBuffers ();

    // Add points from the mixed data to octree
    octree.setInputCloud (cloud);
    octree.addPointsFromInputCloud ();
```

```
        std::vector<int> newPointIdxVector;

        //Get vector of point indices from octree voxels
        //which did not exist in previous buffer
        octree.getPointIndicesFromNewVoxels (newPointIdxVector, 1);

        for (size_t i = 0; i < newPointIdxVector.size(); ++i) {
            fgcloud->push_back(cloud->points[newPointIdxVector[i]]);
        }

        //Filter the fgcloud down
        ror.setInputCloud(fgcloud);
        ror.filter(*fgcloud);

        viewer->updatePointCloud (fgcloud, "Kinect Cloud");
}
```

First, we create a new foreground cloud (fgcloud). We also clear the octree's buffer from the previous iteration. Next, we add the background cloud (bgcloud) to the octree – this is created by stacking successive point clouds (10 in this case), then removing duplicates. This is captured on the first set of iterations after turning on background subtraction. Then, we flip the buffer, keeping the new octree we just created in memory but freeing the input buffer. After that, we insert our new cloud data (cloud) into the octree. PCL has another handy working function called getPointIndicesFromNewVoxels, which provides a vector of indices within the second point cloud (cloud) that are not in the first (bgcloud). Next, we push back the points that are referred to by those indices into our new cloud (fgcloud) and display it. This results in the three output clouds shown in Figures 6-5 (background), 6-6 (full scene), and 6-7 (foreground).

Figure 6-5. The background of the scene, precaptured

Figure 6-6. The full sceneas a point cloud

Figure 6-7. Foreground, after subtraction

Clustering Voxels

Now that we're able to subtract the background and get the foreground only in these images, we're going to need to identify people in the scene. And, in order to do that, we're need to intelligently cluster our voxels together, measure the results, and judge whether or not each result is a person.

There are many clustering algorithms, but we're going to discuss one of the most simple, yet effective, ones—Euclidean clustering. This method is similar to the 2-D flood fill technique. Here's how it works:

1. Create octree O from point cloud P.

2. Create a list of clusters, C, and a cluster, L.

3. For every voxel v in octree O, do the following:

   a. Add v_i to L.

   b. For every voxel v_i in cluster L, do the following:

      • Search in the set O^i_k for neighbors in a sphere with radius r.

      • For every neighbor v^k_i in O^k_i, see if the voxel has already been added to a cluster, and if not, add it to L.

4. When all of the voxels in L have been processed, add L to C, and reset L.

5. Terminate when all voxels v_i in octree 0 have been processed into Ls in C.

Again, PCL provides a function called EuclideanClusterExtraction to save us the pain of having to implement this algorithm ourselves. It is powered by a different tree implementation—a k-d tree. A *k-dimensional (k-d) tree* is another space-partitioning tree; this one is based around splitting along dimensions (octrees are split around a point). K-d trees are also always binary, a property that makes searching for nearest neighbors (or neighbors in a radius) extremely fast.

Clustering in PCL is easy to set up, as shown in Listing 6-6.

Listing 6-6. Clustering Voxels

```
viewer->removeAllShapes();
pcl::PointCloud<pcl::PointXYZRGB>::Ptr clustercloud (new pcl::PointCloud<pcl::PointXYZRGB>);
tree->setInputCloud (fgcloud);
std::vector<pcl::PointIndices> cluster_indices;
ec.setInputCloud( fgcloud );
ec.extract (cluster_indices);
pcl::PointXYZRGB cluster_point;
int numPeople = 0;
int clusterNum = 0;
std::cout << "Number of Clusters: " << cluster_indices.size() << std::endl;
for (std::vector<pcl::PointIndices>::const_iterator it = cluster_indices.begin (); it !=
cluster_indices.end (); ++it)
{
    pcl::PointCloud<pcl::PointXYZRGB>::Ptr cloud_cluster (new
pcl::PointCloud<pcl::PointXYZRGB>);
    float minX(0.0), minY(0.0), minZ(0.0), maxX(0.0), maxY(0.0),     maxZ(0.0);

    std::cout << "Number of indices: " << it->indices.size() << std::endl;
    for (std::vector<int>::const_iterator pit = it->indices.begin ();
pit != it->indices.end (); pit++)
    {
        cluster_point = fgcloud->points[*pit];
        if (clusterNum == 0) cluster_point.r = 255;
        if (clusterNum == 1) cluster_point.g = 255;
        if (clusterNum >= 2) cluster_point.b = 255;
        cloud_cluster->points.push_back (cluster_point);
        if (cluster_point.x < minX)
            minX = cluster_point.x;

        if (cluster_point.y < minY)
            minY = cluster_point.y;

        if (cluster_point.z < minZ)
            minZ = cluster_point.z;

        if (cluster_point.x > maxX)
            maxX = cluster_point.x;

        if (cluster_point.y > maxY)
```

```
        maxY = cluster_point.y;

    if (cluster_point.z > maxZ)
        maxZ = cluster_point.z;
}
```

The KdTree line creates the k-d tree from the input data. A vector called cluster_indices is created to store the indices. setClusterTolerance sets the clustering reach, that is, how far we'll look from each point (the radius in the equation in Listing 6-6). setMinClusterSize and setMaxClusterSize put a floor and ceiling on how many points we'll try to categorize into a single cluster. All three of these values need to be tuned: if they're too small, you'll miss some objects; too large, and you'll categorize multiple objects as one. Our search method is the k-d tree. The extract function fills the vector of PointIndices. When we run this code over our foreground, we end up with the image in Figure 6-8.

Figure 6-8. Clustered voxels

Tracking People and Fitting a Rectangular Prism

Now that we've clustered our data, we can proceed to tracking people and fitting a prism to the data. We're going to use a very simple metric for tracking a person: adults are generally within a range of height and width, as well as being of a certain size. First, we'll judge the height and width of our cluster, and then we'll fit a convex hull to the points to learn what the area is. If a cluster's area is within the metrics, it will be identified as a person, as shown in Listing 6-7.

Listing 6-7. Identifying People

```
//Use the data above to judge height and width of the cluster

if ((abs(maxY-minY) >= 1219.2) && (abs(maxX-minX) >= 101.6)) {
    //taller than 4 foot? Thicker than 4 inches?
    //Draw a rectangle about the detected person
    numPeople++;
    pcl::ModelCoefficients coeffs;
    coeffs.values.push_back ((minX+maxX)/2);//Tx
    coeffs.values.push_back ((minY+maxY)/2);//Ty
    coeffs.values.push_back (maxZ);//Tz
    coeffs.values.push_back (0.0);//Qx
    coeffs.values.push_back (0.0);//Qy
    coeffs.values.push_back (0.0);//Qz
    coeffs.values.push_back (1.0);//Qw
    coeffs.values.push_back (maxX-minX);//width
    coeffs.values.push_back (maxY-minY);//height
    coeffs.values.push_back (1.0);//depth
    viewer->addCube(coeffs, "PersonCube"+boost::lexical_cast<std::string>(numPeople));
}
std::cout << "Cluster spans: " << minX << "," << minY << "," << minZ << " to " << maxX << ","
<< maxY << "," << maxZ << ".  Total size is: " << maxX-minX << "," << maxY-minY << "," <<
maxZ-minZ << std::endl;
clusterNum++;
```

Given these metrics—that a person is larger than 4 feet by 4 inches in the Kinect's view—we get output that looks like Figure 6-9 and Figure 6-10.

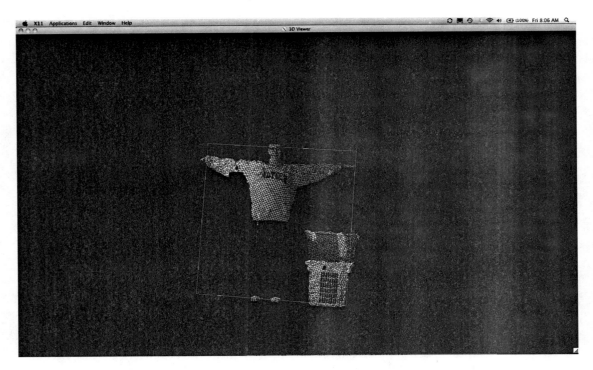

Figure 6-9. *Person tracking*

Figure 6-10. Person tracking

Summary

In this chapter, we delved deep into voxels—why and how to use them—and voxelization. We discussed octrees, background subtraction, clustering, and people tracking. You even had the opportunity to build your very own person tracker.

Point Clouds, Part 1

This chapter brings us into the world of 3-D. You will see what the different ways of representing 3-D data are and how we can transform the 2-D images returned by the Kinect into 3-D points. This chapter will include the following concepts:

- Representing data in 3-D

- Transforming a depth image into a point cloud

- Visualizing a point cloud

- Coloring a point cloud

- Manipulating a point cloud

Let's get started.

Representing Data in 3-D

Before you start producing 3-D models with your Kinect, you must know how to represent and store those models. Most developers are familiar with how 2-D raster images are stored. The image consists of a regular grid of pixels that sample the 2-D space. The image resolution (i.e., how densely the 2-D space is sampled) determines not only the quality of the image but also the processing and memory requirements. Take the Kinect color image, for example (see Figure 7-1). It has VGA resolution (640×480), which means it has 307,200 pixels. But if we double the resolution (1024×960), the pixel count is four times higher. Fortunately, modern computers are able to process images of several megapixels in real time without problems. Yet, when we deal with 3-D data, there are many representation options, each with its own advantages and disadvantages.

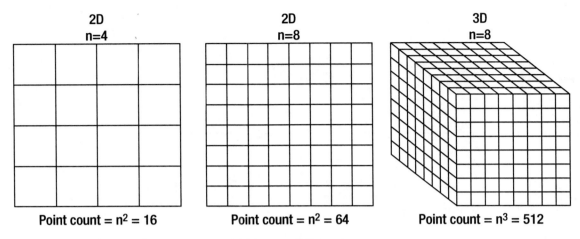

Figure 7-1. Scaling pixel count across resolutions and dimensions

Voxels

A straightforward way to represent a 3-D image would be to create a regular 3-D grid that samples the 3-D space. This creates what is known as a *volumetric image*, and the elements of such a grid are called *voxels*, the 3-D version of pixels. However, when we enter the 3-D world, numbers grow exponentially. For example, extend the low-resolution VGA image along a third dimension ($640 \times 480 \times 480$), and you get 147,456,000 voxels, equivalent to a 140-megapixel image! Volumetric images can be useful and will be explored in a later chapter, but for now, we need a more efficient and compact way of dealing with our 3-D data.

Mesh Models

Mesh models are a very compact way of representing 3-D data and are the representation of choice for computer games and 3-D animation. They model the surface of an object using polygons (usually triangles). Flat areas require only few polygons, while complex areas can use many polygons to represent the surface with high detail. They are the 3-D equivalent of 2-D vector images.

Mesh models are a very flexible and efficient way of representing 3-D data, but they are hard to construct. A 3-D artist selects the vertex positions and normal orientations and determines surface connectivity. However, from the Kinect, we only get the 3-D position and color. We do not (yet) have any information about the orientation of the surface or the connectedness of the pixels in 3-D. Therefore, we need a simpler representation to start.

Point Clouds

A *point cloud* is a collection of unconnected 3-D points. They are called "clouds," because when visualized, the lack of connectivity between points makes them look like they are floating in space. The simplest type of point cloud contains only position information, but each point can also contain other properties (e.g., color and normal orientation). Point clouds can be used as a world model, like in robot navigation to avoid collision with points above the ground plane, or they can be used as an intermediate step before surface reconstruction to produce a mesh model.

Figure 7-2 shows an example of a point cloud. On the left, you see the entire image. On the right, you see a close-up of the raised hand. Even though the hand looks solid from afar, the zoomed view on the right shows the disconnected points.

A point cloud is the natural way of representing the information obtained from the Kinect, because each pixel in the depth image can be transformed into a 3-D point, yet we do not know if it is connected to any other point. We'll be using the Point Cloud Library (PCL) to store the points.

Figure 7-2. Rendering of a point cloud

Creating a Point Cloud with PCL

The basic type of the PCL is the pcl::PointCloud<PointT>. In itself, it is little more than a vector of points but PCL is built around it to provide other services like file I/O, visualization, and normal estimation. Therefore, it is good to start using the PointCloud class from the beginning.

The PointCloud class is templated to be flexible with the type of points we want to store. The simplest point type is pcl::PointXYZ, which consists of only a 3-D position. The type pcl::PointXYZRGB adds RGB color information to each point. The library provides many point types to suit the most common needs, but you are also free to create a custom point class to store other attributes.

If the number of points is known, the point cloud can be preallocated; otherwise, each point can be added with the PointCloud::push_back(PointT) method. The push_back method simply adds each point individually to the cloud while memory is dynamically allocated. Because it is common for many 3-D measurement devices (e.g., time-of-flight cameras, laser scanners, and the Kinect) to return 3-D points in a 2-D grid, the cloud can be indexed linearly as well as two dimensionally. Listing 7-1 shows the creation of a simple random point cloud to illustrate these concepts.

Listing 7-1. Creation of a Point Cloud

```
pcl::PointCloud<pcl::PointXYZ> cloud;

cloud.width = 10;   //Dimensions must be initialized to use 2-D indexing
cloud.height = 10;
cloud.resize(cloud.width*cloud.height);
for(int v=0; v<cloud.height; v++)      //2-D indexing
    for(int ui=0; u<cloud.width; u++) {
        cloud(u,v).x = u;
        cloud(u,v).y = v;
        cloud(u,v).z = rand() / (float) RAND_MAX;   //Assign a random value just for testing
    }
for(unsigned int i=0; i<cloud.size(); i++)     //Linear indexing accesses the same points
    do_something_with_point(cloud[i]);
```

Now that you know how to use PCL to create point clouds, let's look at how we can use the depth image provided by the Kinect to create a 3-D point cloud.

Moving From Depth Map to Point Cloud

The depth data obtained from the Kinect is in the form of a 2-D intensity image. There is clearly some correspondence between true depth (distance from the camera in meters) and the Kinect disparity units used in the depth image, but it might not be obvious how to convert one to the other. This is where the calibration data obtained previously becomes useful.

The conversion from disparity units to meters is done with the following formula:

```
z=1/(value*dc1 + dc2)
```

The conversion is determined by the coefficients dc1 and dc2, which control the scale and offset of the conversion respectively. Ideally, dc1 and dc2 should be calibrated as well, but this requires a different calibration procedure (see www.ee.oulu.fi/~dherrera/kinect/). For now, we will use dc=-0.0030711016 and dc2=3.3309495161. The value 2047 is a special case in the depth image; it indicates an unknown depth.

Once the Z coordinate is determined, we can use the pinhole camera model (see Figure 2-10 in Chapter 2) to obtain the X and Y coordinates. Listing 7-2 shows how to construct a point cloud from a Kinect depth image. In the listing, the depth image is given as an OpenCV image.

Listing 7-2. Transforming a Kinect Depth Image into a Point Cloud

```
pcl::PointCloud<pcl::PointXYZ> cloud;
cv::Mat1s depth_image;
const float dc1= -0.0030711016;
const float dc2=3.3309495161;
float fx_d,fy_d,px_d,py_d;     //From calibration

cloud.width = depth_image.cols;  //Dimensions must be initialized to use 2-D indexing
cloud.height = depth_image.rows;
cloud.resize(cloud.width*cloud.height);
for(int v=0; v< depth_image.rows; v++)     //2-D indexing
    for(int ui=0; u< depth_image.cols; u++) {
```

```
    float z = 1.0f / (depth_image(v,u)*dc1+dc2);

    cloud(u,v).x = z*(u-px_d)/fx_d;
    cloud(u,v).y = z*(v-py_d)/fy_d;
    cloud(u,v).z = z;
}
```

In Listing 7-2, the depth image is represented by an OpenCV image even though OpenKinect puts the image in a raw buffer. However, it is possible (and convenient) to create an OpenCV image and give its internal buffer to OpenKinect. In this way, you can manipulate the image with OpenCV without incurring in any extra memory copying.

■ **Note** OpenCV matrices are indexed with (row, column), but a point cloud is indexed with (column, row).

Coloring a Point Cloud

The point cloud we constructed in the previous section contains only 3-D position information. Since the Kinect also gives us a color image, it would be nice to include color information for each point in the cloud. To do this, the points we obtained from the depth image must be transformed to the coordinate frame of the color camera. Because the cameras have different focal lengths and origins, this cannot be done in 2-D. Fortunately, we get all the needed information from the calibration.

Two steps must be applied to the points to find their color. First, a rigid transformation is applied to transform the points to the coordinate frame of the color camera, and then, the points are projected onto the color camera's image plane.

From Depth to Color Reference Frame

A *rigid transformation* consists of a rotation and a translation. The necessary rotation and translation are named in the calibration data R and T respectively. Expressed as a matrix operation, the transformation we want to apply is simply p_c=R*p_d+T, where R is a 3×3 matrix, and both p_d and T are 3×1 vectors. You will learn more about transformations in the registration section of the next chapter. Figure 7-3 shows how the matrices are used to perform a rotation and a translation. To calculate y_c, the elements of r_2 and p_d are multiplied elementwise and added to t_y.

$$\begin{bmatrix} x_c \\ y_c \\ z_c \end{bmatrix} = \begin{bmatrix} r_{1x} & r_{1y} & r_{1z} \\ r_{2x} & r_{2y} & r_{2z} \\ r_{3x} & r_{3y} & r_{3z} \end{bmatrix} \begin{bmatrix} x_d \\ y_d \\ z_d \end{bmatrix} + \begin{bmatrix} t_x \\ t_y \\ t_z \end{bmatrix}$$

Figure 7-3. Matrix multiplication used for a rigid transformation. The highlighted elements on the right are used to calculate the y coordinate of the tranformed point.

Projecting onto the Color Image Plane

The projection onto the image plane follows the pinhole camera model. However, as mentioned in Chapter 2, imperfections in the camera lens construction and sensor alignment cause distortions from the pinhole model. The pinhole camera model can be extended to include distortion parameters to model radial and tangential distortion. These parameters are also calibrated, but we will ignore them here for clarity. Listing 7-3 shows how to create a colored point cloud.

Listing 7-3. *Creating a Colored Point Cloud*

```
pcl::PointCloud<pcl::PointXYZRGB> cloud;
cv::Mat1s depth_image;
cv::Mat3ub color_image;
const float dc1= -0.0030711016;
const float dc2=3.3309495161;
float fx_d,fy_d,px_d,py_d;    //From calibration
float fx_c,fy_c,px_c,py_c;    //From calibration
cv::Matx33f R;    //From calibration
cv::Matx31f T;    // From calibration

cloud.width = depth_image.cols;  //Dimensions must be initialized to use 2-D indexing
cloud.height = depth_image.rows;
cloud.resize(cloud.width*cloud.height);
for(int v=0; v< depth_image.rows; v++)      //2-D indexing
    for(int ui=0; u< depth_image.cols; u++) {
        cv::Matx31f &Xd = *(cv::Matx31f*)&cloud(u,v).x;    //Access the point as a cv::Matx

        //3-D position
        float z = 1.0f / (depth_image(v,u)*dc1+dc2);
        Xd(0) = z*(u-px_d)/fx_d;
        Xd(1) = z*(v-py_d)/fy_d;
        Xd(2) = z;

        //Project to color image
        cv::Matx31f Xc = R*Xd + T;      //Rigid transformation
        int uc,vc;
        uc = (int)(Xc(0)*fx_rgb/Xc(2) + px_rgb);    //Pinhole projection
        vc = (int)(Xc(1)*fy_rgb/Xc(2) + py_rgb);    //Truncate to nearest neighbor

        //Copy color
        cloud(u,v).r = color_image(vc,uc)[0];
        cloud(u,v).g = color_image(vc,uc)[1];
        cloud(u,v).b = color_image(vc,uc)[2];
    }
```

Visualizing a Point Cloud

Once we have constructed the point cloud, it is useful to visualize it to make sure that we are doing things right. There are two main ways to draw our point cloud: we can give it to PCL for visualization, or we can use OpenGL and draw the point cloud ourselves. Of course, using PCL makes the task much easier, but it might not always be the best option.

Visualizing with PCL

Visualization with PCL requires the Visualization Toolkit (VTK), which can be downloaded from www.vtk.org. If you do not have VTK, PCL will work, but its visualization features will be disabled. The simplest visualization is achieved with only three lines using the pcl::visualization::CloudViewer, as in Listing 7-4.

Listing 7-4. *Simplest Point Cloud Visualization*

```
pcl::PointCloud<pcl::PointXYZRGB> cloud;      //Already populated

pcl::visualization::CloudViewer viewer ("Simple Cloud Viewer");
viewer.showCloud (pcl::PointCloud<pcl::PointXYZRGB>::Ptr(&cloud));      //Ptr must be explicitly
constructed
while(!viewer.wasStopped());      //wasStopped() function processes mouse events for the window
```

PCL has several classes to represent a point depending on what information we want to store with the point. The basic class pcl::PointXYZ contains only position information. Other classes, like pcl::PointXYZI, pcl::PointXYZRGB and pcl::PointNormal, extend this class to include intensity, color, and normal information respectively. The pcl::CloudViewer visualizer only supports clouds with points of type pcl::PointXYZ, pcl::PointXYZI, or pcl::PointXYZRGB. This is good enough for now, but if you want to visualize other point properties (e.g., normal orientation), a more complex visualizer must be used. The pcl::visualization::PCLVisualizer class provides a very flexible interface for implementing complex visualizations. A tutorial of this class is outside of the scope of this book, but luckily, the PCL documentation includes a detailed tutorial on its use.

Visualizing with OpenGL

Depending on the kind of project you want to make, you may need to implement your own visualization. You might want to alter the appearance of the point cloud, or you might want to include the point cloud inside a larger 3-D world. OpenGL is the tool for that.

If you already have a windowed application, you can use OpenGL by itself. If not, the OpenGL Utilities Toolkit (GLUT) makes it easy to set up a window, process events, and render using OpenGL. Follow these steps to set up GLUT and OpenGL:

1. Initialize GLUT/OpenGL.

2. Set up the viewport and camera.

3. Prepare vertex and textures arrays.

4. Render.

Listing 7-5 shows an example initialization and the methods required for rendering.

Listing 7-5. OpenGL Initialization Code

```
int main(int argc, char **argv) {
    glutInit(&argc, argv);      //Initialize GLUT library

    //Create glut window
    glutInitDisplayMode(GLUT_RGBA | GLUT_DOUBLE | GLUT_ALPHA | GLUT_DEPTH);
    glutInitWindowSize(640, 480);
    glutInitWindowPosition(0, 0);
    int main_window = glutCreateWindow("Cloud viewer");

    //Glut callbacks
    glutDisplayFunc(do_glutDisplay);     //These functions are defined by you
    glutIdleFunc(do_glutIdle);                   // to handle the different events
    glutReshapeFunc(do_glutReshape);
    glutKeyboardFunc(do_glutKeyboard);
    glutMotionFunc(do_glutMotion);
    glutMouseFunc(do_glutMouse);

    //Default settings
    glMatrixMode(GL_TEXTURE);
    glLoadIdentity();

    glClearColor(0.0f, 0.0f, 0.0f, 0.0f);
    glEnable(GL_DEPTH_TEST);
    do_glutReshape(Width, Height);

    glutMainLoop();     //Main event loop, will never return
}

// do_glutReshape: called when the window is resized
void do_glutReshape(int Width, int Height) {
    glViewport(0,0,Width,Height);     //Sets the viewport matrix to match the window size
    glMatrixMode(GL_PROJECTION);
    glLoadIdentity();
    gluPerspective(60, 4/3., 0.3, 200);     //Sets up a regular camera in the projection matrix
    glMatrixMode(GL_MODELVIEW);
}

void do_glutDisplay() {
    glClear(GL_COLOR_BUFFER_BIT | GL_DEPTH_BUFFER_BIT);
    //Modelview matrix
    glLoadIdentity();
    glScalef(zoom,zoom,1);     //Matrix transformations to alter where the object is rendered
    glTranslatef(0,0,-3.5);
    glRotatef(rotangles[0], 1,0,0);
    glRotatef(rotangles[1], 0,1,0);
    glTranslatef(0,0,1.5);

    //Show point cloud here
```

```
    glutSwapBuffers();      //Show the newly rendered buffer
}
```

Note OpenGL uses matrices (specifically, GL_PROJECTION and GL_MODELVIEW) to alter the way objects are displayed and to calculate the projection from 3-D world onto the window. These matrices accomplish the same type of rigid transformation and perspective projection as we did manually on the "Coloring a Point Cloud" section.

The actual rendering of the points can be performed in several ways. We can render each point individually, as in Listing 7-6. But this method has a huge overhead when we have many points. OpenGL provides a more efficient way using arrays. Basically, we tell OpenGL where our data is and how many points we wish to draw. The function calls are reduced to only a few, regardless of the point count. See Listing 7-7 for an example of using arrays to draw points.

Listing 7-6. Drawing Points Individually in OpenGL

```
glPointSize(1.0f);
glBegin(GL_POINTS);
for(unsigned int i=0; i<cloud.size(); i++) {
    glColor3ub(cloud[i].r, cloud[i].g, cloud[i].b);
    glVertex3f(cloud[i].x, cloud[i].y, cloud[i].z);
}
glEnd();
```

Listing 7-7. Drawing Points Using Arrays in OpenGL

```
//Set up arrays
unsigned int indices[] = {0,1,2,10,11,12};     //Render only these 6 points
int index_count = 6;
glEnableClientState(GL_VERTEX_ARRAY);
glVertexPointer(3, GL_FLOAT, sizeof(point_cloud[0]), &point_cloud[0].x);
glEnableClientState(GL_COLOR_ARRAY);
glColorPointer(3, GL_FLOAT, sizeof(point_cloud[0]), &point_cloud[0].r);

//Render
glPointSize(1.0f);
glDrawElements(GL_POINTS, index_count, GL_UNSIGNED_INT, indices);
```

Now that you've learned how to create a point cloud, color it, and display it, it is time to put your knowledge to use with a complete application.

EXERCISE 7-1. CLOUD IN THE WIND APPLICATION

Depending on your music taste, you might be familiar with Radiohead's music video "House of Cards." The scenes of the video were scanned using 3-D range equipment, and the video was made entirely of point cloud visualizations. If you haven't seen the video, check it out on YouTube (you can find it under Radiohead's YouTube channel or directly at www.youtube.com/watch?v=8nTFjVm9sTQ). It is a great example of what we can do with point clouds. In this exercise, we'll be imitating some of the effects from the video.

Besides simply plotting the point cloud, Radiohead's video enhanced the visualization by coloring the point cloud, applying TV-like distortions, blowing the points away with virtual wind, and moving the camera smoothly. Of course, what really makes a video great is its artistic touch. In this exercise, we'll be focusing on the technical aspects of how to achieve these effects. It is up to you to decide in which direction the wind blows and how strong, literally.

We will focus on two of the effects of the video: the vertical blue to red coloring and the blowing wind. We will develop an application that draws the point cloud in 3-D and animates wind blowing the points away. The points can be colored using either the blue-red gradient or the Kinect's color image. Because we want to animate the point cloud in real time, we will be using OpenGL to render the point cloud ourselves.

The structure of the application will be similar to what you have seen before. It will have two threads, one for obtaining data from the Kinect, the freenect thread, and one for display, the display thread. This application also requires animation to simulate the wind. Only freenect code will run on the freenect thread, which means that converting the depth image to a point cloud, animation, and display will all run on the display thread.

First, let's go over the libraries we will be using. Listing 7-8 presents the include statements needed for the application. Libfreenect, PThread, PCL, OpenCV, and OpenGL are there, as well as Boost's `date_time` library for the animation. Listing 7-9 shows the code for the freenect thread.

Listing 7-8. Includes for Cloud in the Wind

```
#include <assert.h>
#include <math.h>
#include <iostream>
using std::cout;
using std::endl;

#include <libfreenect.h>

#include <pcl/point_cloud.h>
#include <pcl/point_types.h>
#include <pcl/visualization/cloud_viewer.h>

#include <pthread.h>
#include <opencv2/opencv.hpp>
```

```
#include <GL/glut.h>
#include <GL/gl.h>

#include <boost/date_time.hpp>
using namespace boost::posix_time;

inline float randf() {return rand() / (float)RAND_MAX;}

volatile int die = 0;
pthread_mutex_t mid_buffer_mutex = PTHREAD_MUTEX_INITIALIZER;
pthread_t freenect_thread;
```

Listing 7-9. *Freenect Thread Code*

```
/////////////////////////////////////////////////////////////////
// Freenect code
/////////////////////////////////////////////////////////////////
freenect_context *f_ctx;
freenect_device *f_dev;
freenect_frame_mode video_mode = freenect_find_video_mode(FREENECT_RESOLUTION_MEDIUM,
FREENECT_VIDEO_RGB);
freenect_frame_mode depth_mode = freenect_find_depth_mode(FREENECT_RESOLUTION_MEDIUM,
FREENECT_DEPTH_11BIT);

cv::Mat1s *depth_back, *depth_mid, *depth_front; //Depth image buffers
cv::Mat3b *rgb_back, *rgb_mid, *rgb_front;       //Color image buffers
int got_depth=0, got_rgb=0;

const float calib_fx_d=586.16f; //These constants come from calibration,
const float calib_fy_d=582.73f; //replace with your own
const float calib_px_d=322.30f;
const float calib_py_d=230.07;
const float calib_dc1=-0.002851;
const float calib_dc2=1093.57;
const float calib_fx_rgb=530.11f;
const float calib_fy_rgb=526.85f;
const float calib_px_rgb=311.23f;
const float calib_py_rgb=256.89f;
cv::Matx33f calib_R(  0.99999,   -0.0021409,      0.004993,
                      0.0022251,    0.99985,      -0.016911,
                     -0.0049561,    0.016922,       0.99984);
cv::Matx31f calib_T(  -0.025985,   0.00073534,    -0.003411);

void depth_cb(freenect_device *dev, void *depth, uint32_t timestamp) {
    assert(depth == depth_back->data);
    pthread_mutex_lock(&mid_buffer_mutex);

    //Swap buffers
    cv::Mat1s *temp = depth_back;
    depth_back = depth_mid;
    depth_mid = temp;
```

```
            freenect_set_depth_buffer(dev, depth_back->data);
            got_depth++;

            pthread_mutex_unlock(&mid_buffer_mutex);
    }

    void rgb_cb(freenect_device *dev, void *rgb, uint32_t timestamp) {
            assert(rgb == rgb_back->data);
            pthread_mutex_lock(&mid_buffer_mutex);

            // swap buffers
            cv::Mat3b *temp = rgb_back;
            rgb_back = rgb_mid;
            rgb_mid = temp;

            freenect_set_video_buffer(dev, rgb_back->data);

            got_rgb++;

            pthread_mutex_unlock(&mid_buffer_mutex);
    }

    void *freenect_threadfunc(void *arg) {
            //Init freenect
            if (freenect_init(&f_ctx, NULL) < 0) {
                cout << "freenect_init() failed\n";
                die = 1;
                return NULL;
            }
            freenect_set_log_level(f_ctx, FREENECT_LOG_WARNING);

            int nr_devices = freenect_num_devices (f_ctx);
            cout << "Number of devices found: " << nr_devices << endl;

            if (nr_devices < 1) {
                die = 1;
                return NULL;
            }

            if (freenect_open_device(f_ctx, &f_dev, 0) < 0) {
                cout << "Could not open device\n";
                die = 1;
                return NULL;
            }

            freenect_set_led(f_dev,LED_GREEN);
            freenect_set_depth_callback(f_dev, depth_cb);
            freenect_set_depth_mode(f_dev, depth_mode);
            freenect_set_depth_buffer(f_dev, depth_back->data);
            freenect_set_video_callback(f_dev, rgb_cb);
            freenect_set_video_mode(f_dev, video_mode);
            freenect_set_video_buffer(f_dev, rgb_back->data);
```

```
    freenect_start_depth(f_dev);
    freenect_start_video(f_dev);

    cout << "Kinect streams started\n";

    while (!die && freenect_process_events(f_ctx) >= 0) {
        //Let freenect process events
    }

    cout << "Shutting down Kinect...";

    freenect_stop_depth(f_dev);
    freenect_stop_video(f_dev);

    freenect_close_device(f_dev);
    freenect_shutdown(f_ctx);

    cout << "done!\n";
    return NULL;
}
```

The freenect thread simply receives the images from the Kinect and puts them in the shared middle buffers. You should notice that the buffers for receiving the Kinect data are OpenCV matrices. They are initialized on the `main()` function because both threads use them. Passing the `cv::Mat::data` pointer as the buffer allows us to use OpenCV later to manipulate the images directly. Listing 7-10 shows the code for building the point cloud from the Kinect depth image.

Listing 7-10. *Creating the Point Cloud for the Cloud in the Wind Application*

```
//////////////////////////////////////////////////////////////
// Cloud building
//////////////////////////////////////////////////////////////
typedef struct {
    PCL_ADD_POINT4D;      //PCL adds the x,y,z coordinates padded for SSE alginment
    union {
        struct {
            float tex_u;
            float tex_v;
        };
        float tex_uv[2];
    };
    float intensity;
    float weight;
    float velocity[3];
    EIGEN_MAKE_ALIGNED_OPERATOR_NEW   // make sure our new allocators are aligned
} EIGEN_ALIGN16 TMyPoint;
POINT_CLOUD_REGISTER_POINT_STRUCT (TMyPoint,              // Register our point type with
the PCL
                                   (float, x, x)
                                   (float, y, y)
```

```
                                      (float, z, z)
                                      (float, tex_u, tex_u)
                                      (float, tex_v, tex_v)
                                      (float, intensity, intensity)
                                      (float, weight, weight)
                                      (float[3], velocity, velocity)
    )

pcl::PointCloud<TMyPoint> point_cloud;
cv::Mat_<bool> is_frozen(480,640,false); //Indicates that this point has been touched by
                                         //the wind and will not be refreshed by Kinect
std::vector<unsigned int> valid_indices; //Indices of valid points to render

int cloud_count=0;
float point_cloud_min_x=1e10;       //Min and max values of point cloud
float point_cloud_max_x=-1e10;
float point_cloud_min_y=1e10;
float point_cloud_max_y=-1e10;

//create_point_cloud: creates the point cloud using Kinects color and depth images
void create_point_cloud() {
    float intensity_scale = point_cloud_max_y-point_cloud_min_y;
    valid_indices.clear();

    for(int v=0; v<depth_front->rows; v++)
        for(int u=0; u<depth_front->cols; u++) {
            unsigned int index = v*point_cloud.width+u;
            TMyPoint &p = point_cloud(u,v);
            bool is_valid;

            if(is_frozen(v,u))
                is_valid = true;
            else {
                const short d = depth_front->at<short>(v,u);
                if(d==2047) {
                    is_valid = false;
                    p.x = p.y = p.z = std::numeric_limits<float>::quiet_NaN();
                }
                else {
                    is_valid = true;
                    p.z = 1.0 / (calib_dc1*(d - calib_dc2));
                    p.x = ((u-calib_px_d) / calib_fx_d) * p.z;
                    p.y = ((v-calib_py_d) / calib_fy_d) * p.z;

                    //Project onto color image
                    cv::Matx31f &pm = *(cv::Matx31f*)p.data;
                    cv::Matx31f pc;

                    pc = calib_R*pm+calib_T;

                    float uc,vc;
                    uc = pc(0)*calib_fx_rgb/pc(2) + calib_px_rgb;
```

```
                    vc = pc(1)*calib_fy_rgb/pc(2) + calib_py_rgb;
                    p.tex_u = uc/(float)(rgb_front->cols-1);
                    p.tex_v = vc/(float)(rgb_front->rows-1);

                    p.z = -p.z; //Fix for opengl
                    p.x = p.x;
                    p.y = -p.y;
                    p.intensity = (p.y-point_cloud_min_y) / intensity_scale;
                }
            }

            if(is_valid)
                valid_indices.push_back(index);
        }
    //Calculate point cloud limits for the first 5 frames
    if(cloud_count++ < 5) {
        point_cloud_min_x=point_cloud_min_y=1e10;
        point_cloud_max_x=point_cloud_max_y=-1e10;
        for(unsigned int i=0; i<point_cloud.size(); i++) {
            TMyPoint &p = point_cloud[i];
            if(point_cloud_min_x > p.x)
                point_cloud_min_x = p.x;
            if(point_cloud_max_x < p.x)
                point_cloud_max_x = p.x;
            if(point_cloud_min_y > p.y)
                point_cloud_min_y = p.y;
            if(point_cloud_max_y < p.y)
                point_cloud_max_y = p.y;
        }
    }
}
void show_visualizer() {
    pcl::PointCloud<pcl::PointXYZRGB> cloud2;

    //Copy cloud
    for(unsigned int i=0; i<point_cloud.size(); i++) {
        TMyPoint &p0 = point_cloud[i];
        if(_isnan(p0.x))
            continue;
        pcl::PointXYZRGB p;
        p.x=p0.x; p.y=p0.y; p.z=p0.z;
        int u,v;
        u = (int)(p0.tex_u*639);
        v = (int)(p0.tex_v*479);
        p.r = (*rgb_front)(v,u)[0];
        p.g = (*rgb_front)(v,u)[1];
        p.b = (*rgb_front)(v,u)[2];
        cloud2.push_back(p);
    }

    //Show
    pcl::visualization::CloudViewer viewer ("Simple Cloud Viewer");
```

141

```
        viewer.showCloud (pcl::PointCloud<pcl::PointXYZRGB>::Ptr(&cloud2));
        while(!viewer.wasStopped());
}
```

There are several things to point out in this part of the code. First, notice that we are defining our own
point type TMyPoint. This allows us to store extra information (texture coordinates, weight, and velocity)
that will be used for rendering and animation. We also define a matrix is_frozen that determines whether
a pixel is being animated by the wind. Points that have been touched by the wind (i.e.,
is_frozen(v,u)=true) will not be refreshed from the Kinect data.

The creation of the point cloud is performed in a manner similar to that shown in previous sections. In
addition, we do not reconstruct points that are currently being animated, and pixels with a disparity value
of 2047 are marked as invalid. The 3-D point is also projected onto the color image, but instead of copying
the color, we store the coordinates to use the color image as an OpenGL texture. This is faster, requires
less memory, and can use OpenGL's filtering instead of the nearest neighbor. Finally, we also calculate
how high the point is relative to the height of the scene and store that in the intensity field. This value
will later be used as a one-dimensional texture coordinate into the blue-red gradient texture.

The function show_visualizer() opens PCL's cloud_viewer to inspect the current point cloud. Notice
how we need to create a new point cloud with points of type pcl::PointXYZRGB to be able to use this
viewer.

Listing 7-11. Animation Code for the Cloud in the Wind Application

```
//////////////////////////////////////////////////////////////////////
// Animation
//////////////////////////////////////////////////////////////////////
ptime last_animation_time;
float wind_front_xr = 0.0f;
const float wind_front_velocity=0.03f;

void init_animation() {
    point_cloud.width = 640;
    point_cloud.height = 480;
    point_cloud.is_dense = false;
    point_cloud.resize(640*480);
    for(unsigned int i=0; i<point_cloud.size(); i++) {
        is_frozen[0][i] = false;
        point_cloud[i].x = point_cloud[i].y = point_cloud[i].z =
std::numeric_limits<float>::quiet_NaN();
        point_cloud[i].weight = 1.0 + 10*randf();

        cv::Matx31f &velocity = *(cv::Matx31f*)point_cloud[i].velocity;
        velocity.zeros();
    }
}

cv::Matx31f get_wind_force(TMyPoint &p) {
    float u,v,w;
    float xr = (p.x-point_cloud_min_x) / (point_cloud_max_x-point_cloud_min_x);
//Normalize x
```

```
    u = 1+(randf()-0.5f)*0.6;
    w = (randf()-0.5f)*0.2;
    v = (xr)*sinf(xr*50);
    return cv::Matx31f(u,v,w);
}

void animate() {
    ptime now = microsec_clock::local_time();
    time_duration span = now-last_animation_time;
    float time_ellapsed = span.ticks() / (float)span.ticks_per_second();
    last_animation_time = now;

    wind_front_xr += time_ellapsed*wind_front_velocity; //Wind wall moves from left to
right
    float wind_front_x = wind_front_xr*(point_cloud_max_x-point_cloud_min_x) +
point_cloud_min_x;

    for(unsigned int i=0; i<point_cloud.size(); i++) {
        TMyPoint &p = point_cloud[i];
        if(p.x < wind_front_x || is_frozen[0][i]) {
            is_frozen[0][i] = true;
            cv::Matx31f &velocity = *(cv::Matx31f*)p.velocity;
            cv::Matx31f force=get_wind_force(p);

            velocity += force*(time_ellapsed/p.weight);
            p.x += velocity(0)*time_ellapsed;
            p.y += velocity(1)*time_ellapsed;
            p.z += velocity(2)*time_ellapsed;
        }
    }
}
```

The animation code presented in Listing 7-11 is rather simple. It simulates a wind force sweeping from left to right. The wind starts at the left of the scene and advances with a speed of 0.03m/s to the right. Each point in the cloud is given a random weight at the beginning to give a more natural feel. The force applied by the wind depends on the position of the point and is calculated in the function get_wind_force(). The wind blows to the right, with increasing vertical oscillations, and a small component in a random direction.

As the wind advances, points that are to the left of its planar front are marked as frozen. This means that they will no longer be refreshed from the Kinect data and will only respond to the simulated wind.

The final code, Listing 7-12, presents the OpenGL code for the application. Because the concepts used have already been discussed previously in this chapter, I will only mention briefly each function and its purpose here:

- create_color_map: This creates the blue-to-red gradient texture.

- do_glutIdle: This function is called when GLUT is not handling any other event. We use it to construct the point cloud when new depth data arrives, refresh the color texture when new color data arrives, call the animation function, and terminate the thread when appropriate.

- do_glutDisplay: This function renders the point cloud.

- do_glutKeyboard: We use this one to handle keyboard events.

- do_glutReshape: This updates the OpenGL viewport and projection matrices to match the current window size.

- do_glutMotion: This function captures mouse motion and rotates the scene.

- do_glutMouse: This one captures mouse clicks.

- init_gl: We use this function to initialize the OpenGL and GLUT engines.

- gl_threadfunc: This is the main entry point for the display thread.

- main: This is the main entry point for the application.

Listing 7-12. OpenGL Code for the Cloud in the Wind Application

```
/////////////////////////////////////////////////////////////////
// OpenGL code
/////////////////////////////////////////////////////////////////
int main_window;
float zoom=1;
int mx=-1,my=-1;          // Prevous mouse coordinates
int rotangles[2] = {0}; // Panning angles

const int color_map_size = 256;
unsigned char color_map[3*color_map_size];

//Texture buffer must be power of 2 to avoit OpenGL problems
const int rgb_tex_buffer_width=1024; //Total width of texture buffer
const int rgb_tex_buffer_height=1024;//Total height of texture buffer
int rgb_tex_width; //Real width of texture
int rgb_tex_height; //Real heightof texture
cv::Mat3b rgb_tex(rgb_tex_buffer_height, rgb_tex_buffer_width); //Actual buffer

GLuint gl_rgb_tex;
GLuint gl_colormap_tex;
bool use_rgb_tex=true;

//create_color_map: creates the blue to red gradient 1D texture
void create_color_map() {
    int iteration_size=color_map_size/4;
    int step=color_map_size/iteration_size;
```

```
        int idx = 0;
        int r,g,b, rs,gs,bs;
        int step_count;

        r=0;g=0;b=128;
        for(int k=0; k<5; k++) {
            if(k==0 || k==4) step_count = iteration_size/2; else step_count =
iteration_size;
            switch(k) {
            case 0: rs=0;gs=0;bs=step; break;
            case 1: rs=0;gs=step;bs=0; break;
            case 2:    rs=step;gs=0;bs=-step; break;
            case 3: rs=0;gs=-step;bs=0; break;
            case 4: rs=-step;gs=0;bs=0; break;
            }
            for(int i=0; i<step_count; i++) {
                r+=rs; g+=gs; b+=bs;
                color_map[3*idx+0] = cv::saturate_cast<uchar>(r);
                color_map[3*idx+1] = cv::saturate_cast<uchar>(g);
                color_map[3*idx+2] = cv::saturate_cast<uchar>(b);
                idx++;
            }
        }
}


void do_glutIdle() {
    //Process Kinect data
    if(got_depth > 0 || got_rgb > 0) {
        bool is_depth_new=false, is_rgb_new=false;

        pthread_mutex_lock(&mid_buffer_mutex);
        if(got_depth) {
            //Switch buffers
            cv::Mat1s *temp;

            temp = depth_mid;
            depth_mid = depth_front;
            depth_front = temp;

            got_depth = 0;
            is_depth_new = true;
        }
        if(got_rgb) {
            //Switch buffers
            cv::Mat3b *temp;

            temp = rgb_mid;
            rgb_mid = rgb_front;
            rgb_front = temp;

            got_rgb = 0;
```

```
                is_rgb_new = true;
        }

        pthread_mutex_unlock(&mid_buffer_mutex);

        if(is_depth_new)
            create_point_cloud(); //Create new point cloud from depth map
        if(is_rgb_new)
            cv::resize(*rgb_front, rgb_tex,
    cv::Size(rgb_tex_buffer_width,rgb_tex_buffer_height)); //Make rgb texture

        if ( glutGetWindow() != main_window )
            glutSetWindow(main_window);
        glutPostRedisplay();
    }

    //Animate
    animate();

    //Die?
    if(die) {
        pthread_join(freenect_thread, NULL);
        glutDestroyWindow(main_window);
        pthread_exit(NULL);
    }
}

void do_glutDisplay() {
    glClear(GL_COLOR_BUFFER_BIT | GL_DEPTH_BUFFER_BIT);
    //Modelview matrix
    glLoadIdentity();
    glScalef(zoom,zoom,1);
    glTranslatef(0,0,-3.5);
    glRotatef(rotangles[0], 1,0,0);
    glRotatef(rotangles[1], 0,1,0);
    glTranslatef(0,0,1.5);

    //Show points
    if(valid_indices.size() > 0) {
        glEnableClientState(GL_VERTEX_ARRAY);
        glVertexPointer(3, GL_FLOAT, sizeof(point_cloud[0]), &point_cloud[0].x);

        glEnableClientState(GL_TEXTURE_COORD_ARRAY);
        if(use_rgb_tex) {
            glEnable(GL_TEXTURE_2D);
            glTexImage2D(GL_TEXTURE_2D, 0, 3, rgb_tex_buffer_width,
    rgb_tex_buffer_height, 0, GL_RGB, GL_UNSIGNED_BYTE, rgb_tex.data);
            glTexCoordPointer(2, GL_FLOAT, sizeof(point_cloud[0]),
    &point_cloud[0].tex_uv);
        }
        else {
            glDisable(GL_TEXTURE_2D);
```

```
            glEnable(GL_TEXTURE_1D);
            glTexCoordPointer(1, GL_FLOAT, sizeof(point_cloud[0]),
&point_cloud[0].intensity);
        }

        glPointSize(1.0f);

glDrawElements(GL_POINTS,valid_indices.size(),GL_UNSIGNED_INT,&valid_indices.front());
    }

    glutSwapBuffers();
}

void do_glutKeyboard(unsigned char key, int x, int y) {
    switch(key) {
    case 27:    die = 1; break;
    case 'w':   zoom *= 1.1f; break;
    case 's':   zoom /= 1.1f; break;
    case 't':   use_rgb_tex = !use_rgb_tex; break;
    case 'v':   show_visualizer();
    }
}

void do_glutReshape(int Width, int Height) {
    glViewport(0,0,Width,Height);
    glMatrixMode(GL_PROJECTION);
    glLoadIdentity();
    gluPerspective(60, 4/3., 0.3, 200);
    glMatrixMode(GL_MODELVIEW);
}

void do_glutMotion(int x, int y) {
    if (mx>=0 && my>=0) {
        rotangles[0] += y-my;
        rotangles[1] += x-mx;
    }
    mx = x;
    my = y;
}

void do_glutMouse(int button, int state, int x, int y) {
    if (button == GLUT_LEFT_BUTTON && state == GLUT_DOWN) {
        mx = x;
        my = y;
    }
    if (button == GLUT_LEFT_BUTTON && state == GLUT_UP) {
        mx = -1;
        my = -1;
    }
}

void init_gl(int Width, int Height) {
```

147

```
        //Create glut window
        glutInitDisplayMode(GLUT_RGBA | GLUT_DOUBLE | GLUT_ALPHA | GLUT_DEPTH);
        glutInitWindowSize(640, 480);
        glutInitWindowPosition(0, 0);
        main_window = glutCreateWindow("Cloud in the Wind");

        //Glut callbacks
        glutDisplayFunc(do_glutDisplay);
        glutIdleFunc(do_glutIdle);
        glutReshapeFunc(do_glutReshape);
        glutKeyboardFunc(do_glutKeyboard);
        glutMotionFunc(do_glutMotion);
        glutMouseFunc(do_glutMouse);

        //RGB texture for opengl
        glGenTextures(1, &gl_colormap_tex);
        glBindTexture(GL_TEXTURE_1D, gl_colormap_tex);
        glTexParameteri(GL_TEXTURE_1D, GL_TEXTURE_MIN_FILTER, GL_LINEAR);
        glTexParameteri(GL_TEXTURE_1D, GL_TEXTURE_MAG_FILTER, GL_LINEAR);
        create_color_map();
        glTexImage1D(GL_TEXTURE_1D, 0, 3, color_map_size, 0, GL_RGB, GL_UNSIGNED_BYTE,
color_map);

        rgb_tex_width = 640;
        rgb_tex_height = 640/(float)video_mode.width * video_mode.height;
        glGenTextures(1, &gl_rgb_tex);
        glBindTexture(GL_TEXTURE_2D, gl_rgb_tex);
        glTexParameteri(GL_TEXTURE_2D, GL_TEXTURE_MIN_FILTER, GL_LINEAR);
        glTexParameteri(GL_TEXTURE_2D, GL_TEXTURE_MAG_FILTER, GL_LINEAR);

        //Default settings
        glMatrixMode(GL_TEXTURE);
        glLoadIdentity();

        glClearColor(0.0f, 0.0f, 0.0f, 0.0f);
        glEnable(GL_DEPTH_TEST);
        do_glutReshape(Width, Height);

        last_animation_time = microsec_clock::local_time();
}

void *gl_threadfunc(void *arg) {
        cout << "GL thread started\n";
        cout << "'w' = zoom in, 's' = zoom out, 'v' = launch PCL cloud viewer, 't' = change
texturing mode\n";
        init_gl(640, 480);
        init_animation();

        glutMainLoop();
        return NULL;
}
```

```
/////////////////////////////////////////////////////////////
// Entry point
/////////////////////////////////////////////////////////////
int main(int argc, char **argv)
{
    cout << "Cloud viewer demo application\n";

    glutInit(&argc, argv);

    //Create buffers
    depth_back = new cv::Mat1s(depth_mode.height, depth_mode.width);
    depth_mid = new cv::Mat1s(depth_mode.height, depth_mode.width);
    depth_front = new cv::Mat1s(depth_mode.height, depth_mode.width);
    rgb_back = new cv::Mat3b(video_mode.height, video_mode.width);
    rgb_mid = new cv::Mat3b(video_mode.height, video_mode.width);
    rgb_front = new cv::Mat3b(video_mode.height, video_mode.width);

    //Init threads
    int res = pthread_create(&freenect_thread, NULL, freenect_threadfunc, NULL);
    if (res)
    {
        cout << "pthread_create failed\n";
        return 1;
    }

    // Glut runs on main thread
    gl_threadfunc(NULL);

    return 0;
}
```

Figure 7-4 presents a screenshot of the Cloud in the Wind application. The wind has started to sweep the points on the left.

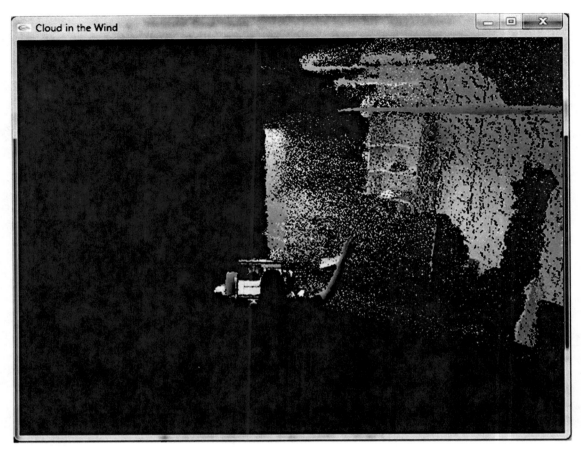

Figure 7-4. Screenshot of the Cloud in the Wind application

Summary

There are many ways of representing 3-D data, including volumetric images, mesh models, and point clouds. If you want to manipulate the data returned by the Kinect in 3-D, the most efficient and straightforward representation is a point cloud. Once the Kinect is calibrated, transforming a depth image to a point cloud and projecting it onto the color image is straightforward. The Point Cloud Library (PCL) provides the necessary classes to store, manipulate, and visualize Kinect data in 3-D. Even for simple projects, using PCL creates almost no overhead and makes debugging and visualizing easy.

Point Clouds, Part 2

This chapter builds on what you learned about point clouds and the Kinect to explore more advanced 3-D concepts. We will extend our methods to handle more than one image at the time. There are many things we could do with a point cloud, and in this chapter, we will explore three areas:

- Registration

- Simultaneous localization and mapping (SLAM)

- Surface reconstruction

Constructing a point cloud from the depth and color images of the Kinect is only the first step. For most applications, we must still do some processing to be able to use the point cloud. So far, the point clouds we've constructed from each frame are independent from each other. It would be very useful if we could relate the different point clouds. For example, if one frame sees the front and left walls of a room and another frame sees the front and right walls, it would be nice to merge the two point clouds and have a single point cloud with all three walls. Finding the transformation that aligns the two point clouds is known as *registration*.

A common use of 3-D information is robot navigation. The structure of the room can be used to plan a route for a robot that avoids obstacles. To plan the route, we need to know the structure of the environment and the robot's location. If we know the environment, we can easily locate the robot. Likewise, if we know the position and trajectory of the robot, we can map the environment's structure. However, estimating both simultaneously is a complicated problem. In robotics, this is called *simultaneous localization and mapping (SLAM)*.

Finally, once we have the environment structure mapped as a point cloud, we need to extract the object surfaces. If you look at the point cloud from far away, it appears to contain solid objects, but once you zoom in, the points are clearly separated and disconnected. If we want to render new views of the environment, we need to interpolate its structure and color between the points. This is achieved by fitting a solid surface to the point cloud and is known as *surface reconstruction*.

Registration

Registration is a fundamental part of computer vision. It is the process of aligning two datasets taken from different coordinate systems. The datasets can be images, in which case registration is performed in 2-D, or they can be 3-D models or point clouds, in which case registration is performed in 3-D. Depending on the task, the estimated transformation can be a simple translation or a complicated nonrigid deformation. A well-known use for registration is in the panorama-making software of consumer cameras, where several images of the scene are stitched together to produce a larger, panoramic image. Registration is used to estimate the relative position of the images and align the overlapping areas.

Registration can be formulated as finding the best transformation that aligns the two datasets. The transformation is applied to the coordinates of the image. In many cases, it is practical to represent the transformation in homogeneous coordinates by a matrix multiplication as follows:

$p' = Tp$

Here, p' is a vector containing the transformed homogeneous coordinates; T is a matrix that represents the transformation, and p represents the original homogeneous coordinates. The structure of the transformation matrix depends on the type of transformation that we want to use.

▪ **Note** Homogeneous coordinates are useful to represent any linear transformation, including translations, with a single matrix multiplication. To convert a point from normal to homogeneous coordinates, simply add an extra element with a value of one (that is, $[x, y]^\mathsf{T} \rightarrow [x, y, 1]^\mathsf{T}$). To convert back, divide by the last element and remove it ($[\omega x, \omega y, \omega]^\mathsf{T} \rightarrow [x, y]^\mathsf{T}$).

2-D Registration

We will start by analyzing 2-D registration of color images. To estimate the transformation between two images, we use point correspondences. Just like when we do object tracking, we detect features, extract descriptors, and then match descriptors between the two images. Figure 8-1 shows two images from an example scene with some of the matched features.

Figure 8-1. Two images of the same scene and some of the matched features

■ **Note** When you use automatic feature matching, some features will be matched incorrectly. These incorrect matches are called *outliers*, because they represent a small percentage of the total matches and produce very high numbers of errors when estimating a transformation. You will learn how to deal with outliers in the section "Robustness to Outliers." For now, we assume all matches are correct.

The simplest type of registration we can do consists of a translation. This is represented by the following transformation of the image coordinates:

$$\begin{bmatrix} x' \\ y' \\ 1 \end{bmatrix} = \begin{bmatrix} 1 & 0 & t_x \\ 0 & 1 & t_y \\ 0 & 0 & 1 \end{bmatrix} \begin{bmatrix} x \\ y \\ 1 \end{bmatrix} = \begin{bmatrix} x + t_x \\ t + t_y \\ 1 \end{bmatrix}$$

To estimate the transformation parameters (t_x and t_y), we use pairs of corresponding points. In theory, for a simple translation, we need only one pair of matched points, but we can use many to improve the estimate. In this case, the least mean square estimate is an intuitive one. The best transformation is obtained from the mean of the point differences.

$$t_x = \sum (x' - x)/N$$
$$t_y = \sum (y' - y)/N$$

Using the transformation, we can project the second image onto the coordinate space of the first image. Figure 8-2 shows the result using the images from Figure 8-1. The base image (left in Figure 8-1) is rendered in black and white. The second image is overlaid on top in full color.

Figure 8-2. *Registration of the images from Figure 8-1 using a translation*

Although a translation is necessary to register images, it is often insufficient. The image can be rotated, scaled, and/or sheared. A more general transformation that includes all these is the affine transformation, which is represented by a 3×3 matrix with a fixed last row of $[0,0,1]$ as in:

$$T_{affine} = \begin{bmatrix} a_{11} & a_{12} & a_{13} \\ a_{21} & a_{22} & a_{23} \\ 0 & 0 & 1 \end{bmatrix}$$

The affine transformation contains six parameters and thus, in theory, requires only three point matches to estimate, but using more points to average out small errors is always good practice. Figure 8-3 presents an image from the same scene as Figure 8-1 presenting rotation and scaling. The overlay of the affine transformed image shows a good match to the original image.

Figure 8-3. *Registration of using an affine transformation*

> ■ **Note** The most general transformation in 2-D is a homography and consists of an arbitrary 3×3 matrix with 11 degrees of freedom (scale is irrelevant in homogenous coordinates). If the scene corresponds to a flat plane, a homography fully describes the relation between any two images of the plane.

3-D Registration

The previous registration examples worked because the scene structure was very close to a plane. If the scene has a 3-D structure, we cannot register it in 2-D without distorting some area of the image. But if we have 3-D information, like a point cloud, we can find a transformation in 3-D that aligns the points.

We will consider the case of a 3-D rigid transformation, which consists of a rotation and a translation. The term "rigid" means that the structure of the object does not change; only the coordinate system is moved and rotated. A rigid transformation between two points in nonhomogenous coordinates is expressed in the following form:

$$p' = Rp + t$$

Here, R is a 3×3 orthonormal matrix that represents the rotation, and t is a 3×1 vector that represents the translation. To estimate the parameters of the transformation, we need 3-D point matches. Given two point clouds, P and P', where every point p_i has a corresponding point p_i', we can easily compute the rigid transformation between them. This can be done by computing first the rotation and then the translation.

To compute the rotation, we first subtract the mean position from each point cloud:

$$m = \frac{1}{N} \sum_{j=1}^{N} p_j$$
$$\bar{p}_i = p_i - m$$

This simplifies the problem, because it is now independent of the translation, and the resulting point clouds are related only by a rotation. We stack the points from each cloud in a matrix $\bar{P}$ of size $3 \times N$, where each column is a point, and multiply them together:

$A = \bar{P}(\bar{P}')^{\mathsf{T}}$

We can extract the rotation from the matrix A by decomposing it through SVD. The SVD decomposition expands a given matrix into three matrices, U, D, and V, of the following form:

$UDV^{\mathsf{T}} = A$

The rotation then is obtained with this formula:

$R = VU^{\mathsf{T}}$

Finally, once we have the rotation, the translation can be obtained with the following equation:

$t = m' - Rm$

■ **Tip** Don't be intimidated by the equations in this section. Sometimes, a little math is necessary to achieve cool things. You will find that these equations' C++ implementations are very straightforward.

Listing 8-1 shows the C++ implementation of the absolute orientation for 3-D points.

Listing 8-1. *Absolute Orientation of Two 3-D Point Clouds*

```
//Inputs
cv::Mat1f X;   //Nx3 matrix of points
cv::Mat1f Y;   //Nx3 matrix of points
//Outputs
cv::Matx33f R; //3x3 rotation matrix (X=R*Y+T)
cv::Matx31f T; //3x1 translation matrix

cv::Matx31f meanX(0,0,0),meanY(0,0,0);
int point_count = X.rows;

//Calculate mean
for(int i=0; i<point_count; i++) {
    meanX(0) += X(i,0);
    meanX(1) += X(i,1);
    meanX(2) += X(i,2);
    meanY(0) += Y(i,0);
    meanY(1) += Y(i,1);
    meanY(2) += Y(i,2);
}
meanX *= 1.0f / point_count;
meanY *= 1.0f / point_count;
```

```
//Subtract mean
for(int i=0; i<point_count; i++) {
    X(i,0) -= meanX(0);
    X(i,1) -= meanX(1);
    X(i,2) -= meanX(2);
    Y(i,0) -= meanY(0);
    Y(i,1) -= meanY(1);
    Y(i,2) -= meanY(2);
}

//Rotation
cv::Mat1f A;
A = Y.t() * X;

cv::SVD svd(A);

cv::Mat1f Rmat;
Rmat = svd.vt.t() * svd.u.t();
Rmat.copyTo(R);

//Translation
T = meanX - R*meanY;
```

> ■ **Note** PCL provides more advanced point cloud registration methods, like the *iterative closest point* algorithm (implemented in the `pcl::IterativeClosestPoint` class). These methods are more time consuming but can also be more accurate.

Robustness to Outliers

Incorrect matches introduce very large errors in the resgistration. They can be regarded as outliers, that is, points that do not match the model. Linear estimation methods, like the ones reviewed in the previous sections, give greater weight to larger errors. This means that outliers, even if they are few, will have a big impact in the estimation. The final transformation will try to accommodate these outliers and will be incorrect.

There are several ways of making the estimation robust to outliers. One option is to constrain the weighting function so that the error stops increasing after a given threshold. This effectively ignores points that are too far away from the estimated solution. However, this makes the estimation nonlinear and requires more complex nonlinear optimization methods.

A very popular and useful algorithm for detecting outliers is Random Sample Consensus (RANSAC). RANSAC relies on the assumption that the outliers constitute a small percentage of the matches. It computes several different solutions and determines which solution has a higher inlier count. The inliers for this solution can then be used with linear methods to estimate the final solution.

To maximize the probability that one of the computed solutions will be close to the real solution, we use the minimium number of points to calculate each of them. This makes it more likely that only inliers will be used to calculate the solution. For example, to estimate a 3-D rigid transformation only three-point matches are needed; therefore, for each iteration, three random matches are used to estimate a transformation.

Inliers are determined by comparing the observed match position with the expected match position given the transformation. If the distance is above a set threshold, the point is considered an outlier.

An implementation of the RANSAC estimation for 3-D absolute orientation is found in Listing 8-2.

Listing 8-2. *RANSAC Estimation of Orientation*

```
void ransac_orientation(const cv::Mat1f &X, const cv::Mat1f &Y, cv::Matx33f &R, cv::Matx31f
&T) {
    const int max_iterations = 200;
    const int min_support = 3;
    const float inlier_error_threshold = 0.2f;

    const int pcount = X.rows;
    cv::RNG rng;
    cv::Mat1f Xk(min_support,3), Yk(min_support,3);
    cv::Matx33f Rk;
    cv::Matx31f Tk;
    std::vector<int> best_inliers;

    for(int k=0; k<max_iterations; k++) {
        //Select random points
        for(int i=0; i<min_support; i++) {
            int idx = rng(pcount);
            Xk(i,0) = X(idx,0);
            Xk(i,1) = X(idx,1);
            Xk(i,2) = X(idx,2);
            Yk(i,0) = Y(idx,0);
            Yk(i,1) = Y(idx,1);
            Yk(i,2) = Y(idx,2);
        }

        //Get orientation
        absolute_orientation(Xk,Yk,Rk,Tk);

        //Get error
        std::vector<int> inliers;
        for(int i=0; i<pcount; i++) {
            float a,b,c,errori;
            cv::Matx31f py,pyy;
            py(0) = Y(i,0);
            py(1) = Y(i,1);
            py(2) = Y(i,2);
```

```
            pyy = Rk*py+T;
            a = pyy(0)-X(i,0);
            b = pyy(1)-X(i,1);
            c = pyy(2)-X(i,2);
            errori = sqrt(a*a+b*b+c*c);
            if(errori < inlier_error_threshold) {
                inliers.push_back(i);
            }
        }

        if(inliers.size() > best_inliers.size()) {
            best_inliers = inliers;
        }
    }
    std::cout << "Inlier count: " << best_inliers.size() << "/" << pcount << "\n";

    //Do final estimation with inliers
    Xk.resize(best_inliers.size());
    Yk.resize(best_inliers.size());
    for(unsigned int i=0; i<best_inliers.size(); i++) {
        int idx = best_inliers[i];
        Xk(i,0) = X(idx,0);
        Xk(i,1) = X(idx,1);
        Xk(i,2) = X(idx,2);
        Yk(i,0) = Y(idx,0);
        Yk(i,1) = Y(idx,1);
        Yk(i,2) = Y(idx,2);
    }
    absolute_orientation(Xk,Yk,R,T);
}
```

Simultaneous Localization and Mapping (SLAM)

SLAM is a very active area in the robotics research community. There are many approaches to it using a great variety of sensors and algorithms. We will focus on the case where only cameras are used, and we will explore how the Kinect simplifies the SLAM problem.

Imagine a robot that is being dragged across a room and whose goal is to return to its initial position afterward. As it is being moved across the room, it has only cameras to learn the path back home. This poses a great challenge, because it must map the structure of the room and determine its changing position simultaneously. As we have seen before, we can obtain feature matches between images as the robot changes position, but the 2-D registration techniques we studied cannot map its 3-D motion across the room.

SLAM Using a Conventional Camera

Using a normal camera, we can track 2-D features across images and try to estimate their 3-D position as the robot moves. We cannot know the 3-D position of a feature from a single frame, thus several frames are needed to estimate the feature positions. But the pose of the camera for each frame is also unknown, which complicates the initial estimation. Although algorithms exist to simultaneously estimate the feature positions and camera poses from a set of images, they are time consuming and susceptible to noise.

Once a valid initial solution has been found, a list of tracked features and their positions is kept along with the robot's estimated trajectory. The algorithm keeps track of uncertainties in the positions— that is, how sure we are of the position of a given feature and the robot. If a feature is observed from many positions, we can estimate its position more accurately. Likewise, if the robot observes many features from a given position, we can estimate its own position more accurately. Finally, as new features are found, they are added to the map and tracked. This will ultimately construct a map with all the observed features and a list of robot positions in the map for each frame.

Position uncertainties are often modeled using a Gaussian model, where the 3-D uncertainty is represented by a 3×3 covariance matrix. This allows us to model high confidence in some direction while having high uncertainty in another direction. This is common with feature positions, because their positions perpendicular to the camera's optical axis can be estimated within pixel accuracy, but the distance from the camera can have high uncertainty.

To incorporate the information from new frames into the current model, many algorithms use a Kalman filter, which predicts the robot's position and updates it along with the features' 3-D positions using the observed 2-D feature positions.

Advantages of Using the Kinect for SLAM

The Kinect simplifies the SLAM problem considerably and is thus a great sensor for robot navigation. The Kinect improves a SLAM algorithm in three ways:

- It is no longer necessary to make an initial guess on the depth of a newly observed feature because the pixels from the Kinect images have depth information.

- The estimation of feature positions requires fewer frames and achieves a more accurate estimate because we have a direct 3-D measurement.

- Instead of adding sparse features to the map, it is possible to add the entire point cloud, creating a dense point cloud that spans the entire room.

There are many ways to use the Kinect for SLAM, like there are many ways to do SLAM without the Kinect. In the next section, we will discuss a specific algorithm to illustrate the idea.

A SLAM Algorithm Using the Kinect

Figure 8-4 shows a diagram of an algorithm that performs SLAM using the Kinect. It uses 2-D feature matching to find the correspondences (although we could include the 3-D information in the feature matching to improve performance) and then uses the 3-D information to construct the map and estimate the camera position.

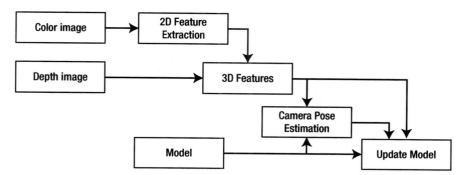

Figure 8-4. *Algorithm for SLAM using Kinect*

The algorithm maintains a model that is constantly updated. The model describes the world and the camera poses at each frame. Like for a conventional camera, the world is described by a list of tracked features. The difference is that when a feature is observed in a Kinect image, it gives us a 3-D position; therefore, we have an accurate map even with a single frame. As the feature is observed in more images, we obtain several measurements of its 3-D position, which can be fused to obtain a better 3-D estimate with less uncertainty.

However, the different measurements are taken from different locations, and since the Kinect returns measurements relative to the camera pose, they must be corrected for camera movement. In other words, each point cloud obtained from the Kinect must be registered in 3-D so that they are all expressed in the same coordinate system. This is called *camera pose estimation*. The features observed in the current frame are matched to the features in the model. Since both have a 3-D position, we can apply the 3-D registration described previously to estimate the camera pose.

■ **Note** One of the biggest improvements over traditional cameras, the ability to estimate the camera pose by using a 3-D registration simplifies the algorithm considerably.

Once we have a list of features for the new frame and an estimate of its camera pose, we can update the model with this new information. Features that were matched can improve the position estimate of the existing features in the model, while unmatched features are added to the model as new tracks.

The model update refines all feature positions and camera poses in the model. This is performed using a nonlinear minimization. The process finds the model parameters that most closely agree with all the observed features. This refinement is especially useful when we return to a part of the scene where we had been before, like when we turn the camera 360 degrees and return to the same initial pose. This is called a *loop closure*, and it helps to stabilize and improve the camera pose estimate.

Real-Time Considerations

SLAM is a very computationally expensive process and implementing it to run in real-time is quite a challenge. The Kinect reduces computational complexity by giving us a 3-D position for new features, but the global refinement stage is still very resource intensive. Depending on the 2-D feature that is used,

161

feature extraction can also be computationally expensive. You are usually left to compromise between speed and robustness.

To have a SLAM system with a global refinement stage and still maintain real-time performance, you need to decouple the refinement stage from the model update. The model is updated after every new frame and the global refinement runs in parallel. In this way, the model is updated in real-time, while the refinement is performed every few seconds.

Special care must be taken to ensure that the nonlinear minimization that performs the global refinement is implemented efficiently. Not all features are visible in all the images, and the algorithm must take this into account to speed up computation. The number of features and camera poses grows very fast, and naïve minimizations that optimize all the parameters with respect to all the observed features can quickly run out of memory, even in modern computers.

Surface Reconstruction

Once we have a point cloud that spans the objects of interest, extracting a surface from it is very useful. A *surface* describes an object's geometry in a more efficient way. Areas with low curvature can be represented with fewer vertices, while highly detailed areas can use more. Moreover, a surface describes the world continuously and is thus able to interpolate between vertices, something that is not easy to do with a point cloud.

A surface is represented by a set of vertices and a set of links. Each vertex contains a 3-D position, color information (either in RGB or in texture coordinates), and a normal direction. The normal direction is not measured by the Kinect and thus must be estimated. The set of links indicates which vertices are connected, defining the topology of the surface. The most common topology is to organize the surface in triangles, with connected triangles sharing two vertices.

Normal Estimation

Using only a point cloud, we can estimate the surface normal direction from the positions of its points. The class pcl::NormalEstimation can be used for this task. It finds the nearest neighbors for each point and fits a plane to them. The normal of the plane is taken as the local surface normal. Because it uses the nearest neighbors independently of their distances, this method can estimate normals with high detail where the point cloud is dense and make a rough estimation where the point cloud is sparse. Listing 8-3 shows how the pcl::NormalEstimation class can be used to estimate normals from a point cloud.

Listing 8-3. *Normal Estimation Using PCL*

```
pcl::PointCloud<pcl::Normal>::Ptr normals (new pcl::PointCloud<pcl::Normal>); //Output normals

//Create search tree
pcl::KdTree<pcl::PointXYZ>::Ptr tree (new pcl::KdTreeFLANN<pcl::PointXYZ>);
tree->setInputCloud (cloud);

//Estimate
pcl::NormalEstimation<pcl::PointXYZ, pcl::Normal> normal_estimator;
normal_estimator.setInputCloud (cloud);
normal_estimator.setSearchMethod (tree);
normal_estimator.setKSearch (20);  //Use 20 nearest neighbors
normal_estimator.compute (*normals);
```

Triangulation of Points

There are several algorithms that can extract a triangle mesh from a point cloud with normal. The main difficulty is to determine which vertices should be connected. Simple distance between vertices is not a good criterion because the scale of the point cloud and its density may vary. Connecting the nearest neighbors is a better way, but care must be taken to avoid self intersections in the surface.

It is out of the scope of this book to go into detail about how these algorithms work. Fortunately, PCL has implementations ready to use. Listing 8-4 shows how to use a simple triangulation method called the greedy projection triangulation.

Listing 8-4. *Normal Estimation Using PCL*

```
// Concatenate the XYZ and normal fields
pcl::PointCloud<pcl::PointNormal>::Ptr cloud_with_normals (new
pcl::PointCloud<pcl::PointNormal>);
pcl::concatenateFields (*cloud, *normals, *cloud_with_normals);

// Create search tree
pcl::KdTree<pcl::PointNormal>::Ptr tree2 (new pcl::KdTreeFLANN<pcl::PointNormal>);
tree2->setInputCloud (cloud_with_normals);

// Initialize objects
pcl::GreedyProjectionTriangulation<pcl::PointNormal> gp3;
pcl::PolygonMesh triangles;

// Set the maximum distance between connected points (maximum edge length)
gp3.setSearchRadius (0.025);

// Set typical values for the parameters
gp3.setMu (2.5);
gp3.setMaximumNearestNeighbors (100);
gp3.setMaximumSurfaceAngle(M_PI/4); // 45 degrees
gp3.setMinimumAngle(M_PI/18); // 10 degrees
gp3.setMaximumAngle(2*M_PI/3); // 120 degrees
gp3.setNormalConsistency(false);

// Get result
gp3.setInputCloud (cloud_with_normals);
gp3.setSearchMethod (tree2);
gp3.reconstruct (triangles);
```

EXERCISE 8-1. A SIMPLE KINECT SLAM

As a final exercise for this chapter, we will be implementing a simple SLAM application using the Kinect. The application will be able to track the camera position in 3-D as you move the Kinect freely with your hand. The point clouds from different points of view will be registered, producing one common point cloud that covers the entire room. A sample screenshot of a point cloud spanning several views is shown in Figure 8-5.

Figure 8-5. *Screenshot from the Simple Kinect SLAM application*

The structure of the algorithm was presented in Figure 8-4 and discussed in the previous sections. To keep things simple, we have omitted the global refinement stage. This means that consecutive frames are registered to each other using only 3-D registration and without the nonlinear minimization and loop closure steps. Even though this leads to some drifting of the camera position over time, we can get a very accurate reconstruction thanks to the 3-D information from the Kinect.

For this application, we will use C++ classes because a regular C program begins to get very cluttered as the complexity increases. This allows us to encapsulate the different modules of the application. To have a responsive UI, each module will have its own thread of execution (three in total). This allows the SLAM code to run independently from the rendering and image acquisition modules. The class structure of the application is outlined in Figure 8-6.

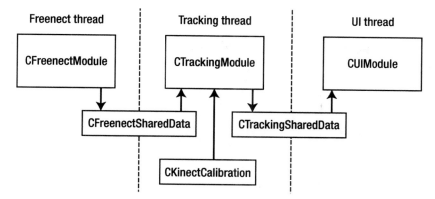

Figure 8-6 *Main classes of the Simple Kinect SLAM application*

Following are the main classes and a brief description of each class:

- CFreenectModule: Handles all communication with the Freenect library. Acquires color and depth images from the Kinect and offers them to the other modules.

- CFreenectSharedData: Stores the new images obtained from the Kinect. Shared between the Freenect and tracking modules.

- CTrackingModule: Executes the SLAM algorithm. Extracts 2-D features, builds point clouds, and registers the point clouds.

- CTrackingShareData: Stores the current model and latest tracked features. Shared between the tracking and UI modules.

- CUIModule: Handles all communication with the user. Uses the GLUT framework to receive user input and renders the aggregated point cloud with OpenGL.

- CKinectCalibration: Encapsulates the calibration parameters of the Kinect and performs conversion between image coordinates and world coordinates.

Most of the code for these modules has already been examined in the previous chapters. For example, the functions of the Freenect module are almost the same as those from the previous chapter, but refactored into a class. We will not discuss these functions again here. Also, because of its multithreaded nature, the code must deal with mutexes to avoid race conditions. *Race conditions* happen when two threads attempt to modify the same variables at the same time. Mutexes are used to synchronize access to these shared variables and allow only one thread at a time to modify them. But we'll skip these details and move right into the interesting parts. You can download the complete code from the accompanying web site of this book.

The Model

The most important part of the application is contained in the tracking module. The model we are using for our SLAM application is contained in the CTrackingSharedData class and is shown in Listing 8-5.

Listing 8-5. Classes Containing the SLAM Model

```
class CFeatureTrack {
public:
    cv::Point2f base_position;
    cv::Point2f active_position;
    cv::Mat1f descriptor;
    int missed_frames;
};

class CTrackedView {
public:
    std::vector<pcl::PointXYZRGB> cloud;
    cv::Matx33f R;
    cv::Matx31f T;
};

class CTrackingSharedData {
public:
    //Commands
    bool is_data_new;    //True if this class has new data that should be rendered
    bool is_tracking_enabled; //True if the tracking thread should process images

    //Base image
    cv::Matx33f base_R;
    cv::Matx31f base_T;
    cv::Mat3b *base_rgb;
    cv::Mat3f *base_pointmap;

    //Last tracked image
    cv::Mat3b *active_rgb;
    cv::Mat1s *active_depth;

    //Model
    std::list<CFeatureTrack> tracks; //Tracked features since last base frame
    std::vector<CTrackedView> views; //All registered views

    CTrackingSharedData();
    ~CTrackingSharedData();
}
```

The CFeatureTrack class holds information about a feature being tracked. Since the feature can be tracked across several images before a frame is registered and added to the model, base_position is the 2-D position where the frame was observed in the last registered frame, and active_position holds the position in the most recent frame. The descriptor matrix holds the SURF descriptor of the feature. It is updated after each successful match to reflect the latest appearance. Finally, if the feature is not matched in the latest image, its missed_frames counter is increased. Frames that miss too many frames are deleted.

The CTrackingView class holds a point cloud and the camera pose (R and T) for a registered view. The camera pose for the first frame is arbitrary and is chosen as the identity matrix. We don't store features for every frame because we need features and descriptors only for the last registered frame. All registered frames before that are stored in the CTrackingSharedData::views vector for display purposes.

The elements with the prefix base contain information about the last registered view, while those with the prefix active contain the last received image. The vector of feature tracks CTrackingSharedData::tracks holds features extracted from the base_rgb image and tracked along images received after it.

2-D Feature Extraction

We use SURF to extract features from the color image, as shown in Listing 8-6.

Listing 8-6. *Extracting Features Using SURF*

```
cv::Mat3b *rgb_buffer; //Pointer to the color image from the Kinect
std::vector<cv::KeyPoint> feature_points; //Extracted feature points
std::vector<float> feature_descriptors; //Descriptor data returned by SURF

cv::Mat1b gray_img;
cv::cvtColor(*rgb_buffer,gray_img,CV_RGB2GRAY,1); //Convert the image to gray level

cv::SURF mysurf(100,4,1,false,false);
mysurf(gray_img, cv::Mat(), feature_points, feature_descriptors); //Extract SURF
features

//Reshape the continuous descriptor vector into a matrix
cv::Mat1f feature_descriptors_mat(feature_points.size(), mysurf.descriptorSize(),
&feature_descriptors[0], sizeof(float)*mysurf.descriptorSize());
```

Once features are extracted for a frame, they can be matched to the features being tracked from the model. OpenCV has an infrastructure in place to provide feature matching. We use here the cv::FlannBasedMatcher. FLANN is short for "Fast Library for Approximate Nearest Network," and it is an efficient algorithm to find nearest neighbors. Using this class, we can find the two closest matching

167

descriptors. If their matching scores are too similar, we consider the match ambiguous and discard it. However, if one match is considerably better than all others, we keep it. This provides very good matching performance and scalability. The implementation of this is found in Listing 8-7.

Listing 8-7. *Matching SURF Features*

```
void CTrackingModule::match_features(const cv::Mat1f &new_descriptors, std::vector<int>
&match_idx) {
    cv::FlannBasedMatcher matcher;
    std::vector<cv::Mat> train_vector;
    std::vector<std::vector<cv::DMatch>> matches;

    train_vector.push_back(new_descriptors);
    matcher.add(train_vector);

    match_idx.resize(shared.tracks.size());

    std::list<CFeatureTrack>::iterator track_it;
    int i;
    for(i=0,track_it=shared.tracks.begin(); track_it!=shared.tracks.end();
i++,track_it++) {
        matcher.knnMatch(track_it->descriptor, matches, 2);
        float best_dist = matches[0][0].distance;
        float next_dist = matches[0][1].distance;

        if(best_dist < 0.6*next_dist)
            match_idx[i] = matches[0][0].trainIdx;
        else
            match_idx[i] = -1;
    }
}
```

Registering every frame we obtain and adding all its points to the model can make the aggregated point cloud grow disproportionally large and redundant. For example, if the Kinect stops moving and looks at the same scene for several frames, there is no need to add the same points over and over again. To avoid this, we calculate how much the features have moved between frames and register the frame only if the movement is above a given threshold. This can be done very efficiently in 2-D. The function `CTrackingModule::get_median_feature_movement()`, presented in Listing 8-8, returns the median movement that is checked before performing registration.

Listing 8-8. *Computing the Median Feature Movement in 2-D*

```
float CTrackingModule::get_median_feature_movement() {
    std::vector<float> vals;
```

```
        float sum=0;
        int count=0;

        std::list<CFeatureTrack>::iterator track_it;
        for(track_it=shared.tracks.begin(); track_it!=shared.tracks.end(); track_it++) {
            if(track_it->missed_frames == 0) {
                vals.push_back(fabs(track_it->base_position.x - track_it->active_position.x)
+
                    fabs(track_it->base_position.y - track_it->active_position.y));
            }
        }

        if(vals.empty())
            return 0;
        else {
            int n = vals.size()/2;
            std::nth_element(vals.begin(),vals.begin()+n,vals.end());
            return vals[n];
        }
}
```

Constructing the Point Map

Chapter 7 introduced a straightforward way of coloring the point cloud obtained from the depth image. However, that method is not entirely accurate because some points observed from the depth image are not visible from the color camera. Points that are occluded in the color camera should be removed from the point cloud because we do not know their colors.

To accomplish this, we construct a point map. The point map is an image with the same coordinates as the color image that stores 3-D points. As 3-D points are reconstructed from the depth image, they are projected onto the color image and added to the point map at that location. If the point map already has a point for that pixel, only the closest point is retained. This discards occluded points. The point cloud is then constructed only from points in the point map. Listing 8-9 presents the functions used to build the point map from the depth image and then the point cloud from the point map.

Listing 8-9. Construction of a Point Map and a Point Cloud

```
void CTrackingModule::compute_pointmap(const cv::Mat1s &depth, cv::Mat3f &pointmap) {
    pointmap = cv::Mat3f::zeros(480,640);

    for(int v=0; v<depth.rows; v++)
        for(int u=0; u<depth.cols; u++) {
            const short d = depth(v,u);
            if(d==2047)
```

```
                continue;

            cv::Matx31f p;
            float uc,vc;
            int uci,vci;
            calib.disparity2point(u,v,d,p);
            calib.point2rgb(p,uc,vc);
            uci = (int)uc+0.5f;
            vci = (int)vc+0.5f;
            if(uci<0 || uci>=pointmap.cols || vci<0 || vci>=pointmap.rows)
                continue;

            cv::Vec3f &point = pointmap(vci,uci);
            if(point(2) == 0 || point(2) > p(2)) {
                point(0) = p(0);
                point(1) = p(1);
                point(2) = p(2);
            }
        }
}

void CTrackingModule::cloud_from_pointmap(const cv::Mat3b &rgb, const cv::Mat3f
&pointmap, std::vector<pcl::PointXYZRGB> &cloud) {
    for(int v=0; v<pointmap.rows; v++)
        for(int u=0; u<pointmap.cols; u++) {
            const cv::Vec3f &pm = pointmap(v,u);
            if(pm(2)==0)
                continue;
            const cv::Vec3b &color = rgb(v,u);

            pcl::PointXYZRGB p;
            p.x = pm(0);
            p.y = pm(1);
            p.z = pm(2);
            p.r = color(2);
            p.g = color(1);
            p.b = color(0);

            cloud.push_back(p);
        }
}
```

Having a point map, we can readily transform a 2-D feature into 3-D; we just have to look up the point stored on the corresponding pixel. We now have a set of matched 3-D features and are ready for the camera pose estimation.

Camera Pose Estimation

The function CTrackingModule::transformation_from_tracks(), shown in Listing 8-10, estimates the camera pose from the feature tracks in the model. First, the 3-D features are stacked into two Nx3 matrices. The 3-D positions from the last registered view are stored in the matrix X, while those from the latest frame are stored in matrix Y.

Listing 8-10. *Function CTrackingModule::transformation_from_tracks()*

```
void CTrackingModule::transformation_from_tracks(const cv::Mat3f &active_pointmap,
cv::Matx33f &R, cv::Matx31f &T) {
    std::list<CFeatureTrack>::iterator track_it;

    cv::Mat1f X(0,3), Y(0,3);
    X.reserve(shared.tracks.size());
    Y.reserve(shared.tracks.size());

    for(track_it=shared.tracks.begin(); track_it!=shared.tracks.end(); track_it++) {
        if(track_it->missed_frames!=0)
            continue;

        int ub=round<int>(track_it->base_position.x),vb=round<int>(track_it-
>base_position.y);
        cv::Vec3f &base_point = (*shared.base_pointmap)(vb,ub);
        if(base_point(2)==0)
            continue;

        int ua=round<int>(track_it->active_position.x),va=round<int>(track_it-
>active_position.y);
        const cv::Vec3f &active_point = active_pointmap(va,ua);
        if(active_point(2)==0)
            continue;

        //Add to matrices
        int i=X.rows;
        X.resize(i+1);
        X(i,0) = base_point(0);
        X(i,1) = base_point(1);
        X(i,2) = base_point(2);

        Y.resize(i+1);
        Y(i,0) = active_point(0);
        Y(i,1) = active_point(1);
        Y(i,2) = active_point(2);
    }
```

171

```
    ransac_orientation(X,Y,R,T);
}
```

Once the matrices have been constructed `CTrackingModule::ransac_orientation()` is called to perform an absolute 3-D orientation that is robust to ouliers (see Listing 8-2).

After we have registered the view, it is simply added to the `CTrackingSharedData::views` vector. The UI module will render all point clouds in a common coordinate system. Figure 8-6 shows two views of a 360-degree reconstruction using 30 registered frames. Notice how the overall quality of the registration is very good, although some misalignments are visible. The start and end point of the reconstruction is located on the blue curtains in the corner. Notice how the reconstructed curtains from the first and last views do not do not exactly match. This is because of the accumulated drift in camera pose and requires global refinement to be corrected.

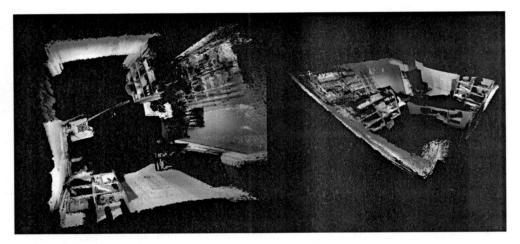

Figure 8-6. *Main classes of the Simple Kinect SLAM application*

Summary

Many applications have a continuous stream of images available to them, especially when using the Kinect. To use more than one image, it is necessary to establish a relation between them. 2-D image registration techniques can be useful under certain camera motions. However, if the camera moves freely in 3-D, a SLAM architecture is necessary. SLAM maps the environment and estimates the camera poses simultaneously. The Kinect is an excellent tool for SLAM, as it simplifies the algorithm considerably and improves its quality. Once the point cloud obtained covers all necessary areas of the scene with sufficient density, surface reconstruction methods can be used to extract a mesh surface from the point cloud.

Object Modeling and Detection

The acquisition and recognition of 3-D models of objects are among of the most active areas of the computer vision community. With the wide availability of a low-cost 3-D camera, a new range of applications can be considered—for example, personal robotics or augmented reality. In this chapter, we make a step in this direction by showing how 3-D models of everyday objects can be acquired much more easily than before using a Kinect. We also consider the further detection of these modeled objects in new scenes, enabling new, object-based interactions with your personal computer.

The automatic acquisition of 3-D models of everyday objects has many potential applications. Just to mention a few examples, you could do landscaping with your own objects; shops could easily create 3-D online galleries of their products; a personal robot could detect and manipulate the items in your kitchen; you could insert your favorite objects to video games; or you could even duplicate existing objects using a plastic 3-D printer such as the MakerBot Thing-O-Matic! There are still some challenging issues to be solved to reach the level of robustness and accuracy required for these applications, but you will see in this chapter that the Kinect makes it much easier and closer than ever.

The chapter is divided in three parts. We will first explain how rough models can be acquired using a single Kinect image, and thus a single viewpoint. Although a single viewpoint is effective for acquiring information about a scene at a glance, if more details are required, it is necessary to merge the information coming from several viewpoints. You will learn to build a support with markers to estimate the pose of the Kinect at each viewpoint. Finally, we will show how objects lying on a table can be recognized and their locations estimated by using some global descriptors.

Note What is an object model? The objective of object modeling (also called object reconstruction) is to gather and represent the information associated with a real-world object. This usually takes the form of a triangulated mesh, sometimes simplified to get a computer-assisted design (CAD) model, or it can simply be a point cloud of the different views. A set of color images can also be associated to the model if texture-based recognition is considered.

Acquiring an Object Model Using a Single Kinect Image

Since the Kinect gives a dense depth map of a scene, a lot of information can be extracted from a single image. This is very convenient because it avoids the need to move the object or the camera to get a complete set of viewpoints. Of course, since only a small part is visible in a single image, this is adapted only to situations where we have some *apriori* knowledge about the object geometry and where a precise mesh is not required.

Tabletop Object Detector

The first issue when modelingis to determine the object of interest in the scene. Indeed, in Figure 9-1, the object could be the table, the floor, a foot of the table, or any of the items lying on the table. This differentiation is called the segmentation process and consists of separating in the scene the objects (foreground) from the rest (background). When using a Kinect, a very popular way of doing this is to assume that the objects are lying on a flat table, similar to our sample scene in Figure 9-1. As you will see, a flat 3-D plane can be extracted quite precisely and robustly using depth data.

Once the table plane has been estimated, the objects of interest can be segmented by clustering the points lying over the table. Then, the modeling itself will be applied to each of these clusters to acquire a model of each object.

Figure 9-1. Left: A sample scene, with a white table and six objects with different shapes. Right: The corresponding depth image output by the Kinect, thresholded at 1.6 meters (4.6 feet).

Extracting the Table Plane

A plane can be defined using a parametric model with four components (a, b, c, d), using the equation ax+by+cz+d = 0. The objective of this step is to find the parameters that best fit into the depth image to extract the dominant plane of the scene. Since many points are not part of the table, classical fitting techniques such as least squares will not perform well. To fit parametric models to data with many outliers, the Random Sample Consensus (RANSAC) family of algorithms has proven to be very successful.

Noting that three 3-D points are sufficient to estimate a plane equation, the idea of RANSAC is to iteratively draw three random points and check how many of the remaining points lie on the estimated plane (inliers), using some distance threshold. The plane that got the highest number of inliers is finally kept, and its equation is refined using least squares fitting with the inliers. If there is a dominant plane, it will be detected in a few iterations, since there is a high probability that three random points belong to it. Figure 9-2 illustrates the algorithm for a 2-D line fitting.

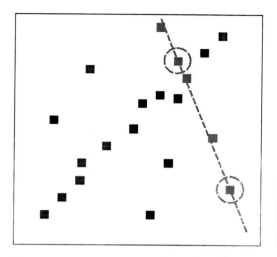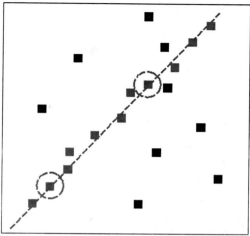

Figure 9-2. Illustration of RANSAC to fit a parametric line to a set of noisy points. The left figure shows an iteration where two outliers were randomly selected as the initial points to initialize the line parameters. This results in only five inliers. When two points on the dominant line are randomly selected (iteration on the right), a higher number of inliers are detected, and thus this line will be returned as the best fit by the algorithm.

RANSAC fitting is implemented in PCL for a number of parametric models, including 3-D planes. In the Listing 9-1, we use a more robust plane fitting that also ensures that the normals of the inliers are consistent.

Listing 9-1. Table Plane Detection

```
//Inputs
pcl::PointCloud<pcl::PointXYZ>::Ptr cloud; // input cloud
pcl::PointCloud<pcl::Normal>cloud_normals; // normals of the point cloud
//Outputs
pcl::PointIndices table_inliers; // point indices belonging to the table plane
pcl::ModelCoefficients table_coefficients;

pcl::SACSegmentationFromNormals<pcl::PointXYZ,pcl::Normal> segmentor;
segmentor.setOptimizeCoefficients(true);
segmentor.setModelType(pcl::SACMODEL_NORMAL_PLANE);
segmentor.setMethodType(pcl::SAC_RANSAC);
segmentor.setProbability(0.99);
// Points at less than 1cm over the plane are part of the table
segmentor.setDistanceThreshold (0.01);

segmentor.setInputCloud (cloud_downsampled_);
segmentor.setInputNormals (cloud_normals_);
```

```
segmentor.segment (table_inliers, table_coefficients);
```

Using this detector, we can get the equation of the dominant plane in the point cloud, along with the list of point indices belonging to it. The output of this code is shown for our table scene in Figure 9-3. Note that this task, which turns out of be quite easy with a Kinect, would be very hard to perform using a regular webcam, since it is not possible to extract 3-D points in the scene.

■ **Note** Regarding point indices, it is sometimes necessary to work with only a subset of a point cloud. Instead of systematically creating a new point cloud and copying the subset of points into it, many PCL functions optionally take a vector of indices (pcl::PointIndices) that specifies the set of points to be considered in the algorithm. This mechanism is useful to save memory and processing time.

Figure 9-3. *Left:Inliers detected as part of the table plane. Right: Estimated plane cropped to the table border using the convex hull of the inliers.*

Keeping Only Points Lying on the Table

The RANSAC fitting returns a 3-D, infinite plane equation. To determine the location of the candidate objects, we need to limit the working area to the actual boundaries of the table.These can be efficiently approximated by projecting the table inliers onto the plane using the estimated equation and computing its convex hull, as shown in Figure 9-3. This hull delimits the area where the objects can be located. To filter out points that do not lie over this area, the second step is to build a 3-D prism over the hulland eliminate points that are not inside the prism using a geometric check. Here again, as shown in Listing 9-2, PCL is very helpful.

Listing 9-2. *Removing the Background*

```
//Inputs
pcl::PointCloud<pcl::PointXYZ>::Ptr cloud; // input cloud
pcl::PointIndices table_inliers; // point indices belonging to the table plane
pcl::ModelCoefficients table_coefficients; // coefficients (a,b,c,d) of
                                           // ax + by + cz +d = 0

//Outputs
pcl::PointCloud<pcl::PointXYZ> cloud_objects; // points belonging to objects
```

```
// Project the table inliers to the estimated plane
pcl::PointCloud<pcl::PointXYZ> table_projected;

pcl::ProjectInliers<pcl::PointXYZ> proj;
proj.setInputCloud(cloud);
proj.setIndices(table_inliers);
proj.setModelCoefficients(table_coefficients);
proj.filter(table_projected);

// Estimate the convex hull of the projected points
pcl::PointCloud<pcl::PointXYZ> table_hull;
pcl::ConvexHull<pcl::PointXYZ> hull;
hull.setInputCloud(table_projected.makeShared());
hull.reconstruct(table_hull);

// Determine the points lying in the prism

pcl::PointIndices object_indices; // Points lying over the table
pcl::ExtractPolygonalPrismData<pcl::PointXYZ> prism;
prism.setHeightLimits(0.01,0.5); // object must lie between 1cm and 50cm
                                 // over the plane.
prism.setInputCloud(cloud);
prism.setInputPlanarHull(table_hull.makeShared());
prism.segment(object_indices);

// Extract the point cloud corresponding to the extracted indices.
pcl::ExtractIndices<Point>extract_object_indices;
extract_object_indices.setInputCloud(cloud);
extract_object_indices.setIndices(
    boost::make_shared<const pcl::PointIndices>(object_indices));
extract_object_indices.filter(cloud_objects);
```

Now, we have a point cloud containing all the points that corresponds to objects lying on the table. The next step is to separate them into distinct objects.

Extracting Individual Object Clusters

Since each object correspond to a compact cloud, and assuming that each object is far enough from others, clustering techniques can be applied to obtain a separate point cloud for each object. Since we are in a metric space, this can be done with asimple Euclidian clustering, as shown in Listing 9-3. The idea is to connect points that are close enough from each other, according to some threshold (we chose 5 centimeters here). This leads to a disjoint set of connected points, where two clusters are disconnected if the minimal distance between their respective points is greater than the threshold. The output of this clustering step on our sample scene is shown in Figure 9-4.

Listing 9-3. *Extracting Object Clusters*

```
// Input
pcl::PointCloud<pcl::PointXYZ>::Ptr cloud_objects; // points belonging to
                                                   // objects

// Output
// vector of objects, one point cloud per object
std::vector<pcl::PointCloud<pcl::PointXYZ>::Ptr> objects;

pcl::EuclideanClusterExtraction<pcl::PointXYZ> cluster;
cluster.setInputCloud(cloud_objects);
std::vector<pcl::PointIndices> object_clusters;
cluster.extract(object_clusters);

pcl::ExtractIndices<pcl::PointXYZ> extract_object_indices;
for(inti=0; i < object_clusters.size(); ++i)
{
    pcl::PointCloud<pcl::PointXYZ>::Ptr object_cloud;
    object_cloud = new pcl::PointCloud<pcl::PointXYZ>;
    extract_object_indices.setInputCloud(cloud);
    extract_object_indices.setIndices(
        boost::make_shared<constpcl::PointIndices>(object_clusters[i]));
    extract_object_indices.filter(object_cloud);
    objects.push_back(object_cloud);
}
```

Figure 9-4. *Left: Object points detected as lying over the table. Right: After clustering, each object has its own point cloud, materialized by a different color.*

Fitting a Parametric Model to a Point Cloud

We now have a segmented point cloud, with one individual point cloud for each object. For many applications, this is not satisfactory, and a full, triangulated 3-D mesh of the objects is desired. If we know beforehand a parametric shape model for a particular object, we can rely on the same model fitting that we used for the table detection to build an approximated mesh of the object. We give an example here for spherical objects, which can be modeled by a sphere. The parameters to be estimated are then the 3-D location of the sphere center and its radius. Listing 9-4 also uses RANSAC to estimate the best parameters, and a sample result is presented in Figure 9-5.

Listing 9-4. *Fitting a Spherical Model to a Point Cloud*

```
// Input
pcl::PointCloud<Point>cloud; // the object point cloud
// Output
Eigen::VectorXf sphere_parameters; // 3-Dcenter and radius of the fitted
                                   // sphere (four parameters)

pcl::SampleConsensusModelSphere<pcl::PointXYZ>::Ptr sphere_model;
sphere_model = new pcl::SampleConsensusModelSphere<pcl::PointXYZ>(cloud));
pcl::RansacSampleConsensus<pcl::PointXYZ> ransac(sphere_model);
ransac.setDistanceThreshold(.004); // 4 millimeters threshold for inliers
ransac.computeModel();
ransac.getModelCoefficients(sphere_parameters);
```

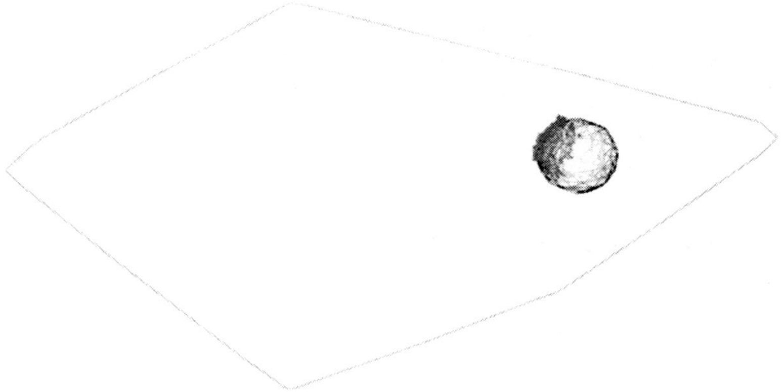

Figure 9-5. *Fitting a sphere to the tennis ball point cloud. The center and radius of the sphere are optimized through RANSAC.*

Building a 3-DModel by Extrusion

If we do not know *a priori* a good parametric model of the shape of an object, we have to relax the hypotheses on the object shape. In this section, we propose a technique to model objects that can be approximated by an extrusion of their topviews. Since this is the case of many manufactured objects, this technique can be used in many cases to get a good approximation of the object mesh.

This will be implemented through an extruder class, which takes an object point cloud as an input and outputs a triangulated mesh. The algorithm is divided into several steps:

1. Build a voxel representation of the object point cloud.

2. Fill the voxels lying between the cloud and the table plane.

3. Build a point cloud representing the surface of the extruded mesh.

4. Build the final triangulated mesh.

Listing 9-5 provides the interface of the class.

Listing 9-5. *The Extruder Class*

```
class Extruder
{
    // Handy typedefs
    typedef pcl::PointCloud<pcl::PointXYZ> PointCloudType;
    typedef PointCloudType::Ptr PointCloudPtr;
    typedef PointCloudType::ConstPtr PointCloudConstPtr;

    // Represents a discrete cell
    struct Voxel;

public:
    Extruder(){}

    void setInputCloud(PointCloudConstPtr cloud)  { input_cloud_ = cloud; }
    void setTablePlane(const Eigen::Vector4f& abcd) { plane_coeffs_ = abcd; }
    void setVoxelSize(float size) { voxel_size_ = size; }

    // Main computing function to build the mesh.
    void compute(pcl::PolygonMesh& output_mesh);

private:
    // Fill the initial voxel map with the given cloud.
    void buildInitialVoxelSet();

    // Fill missing voxels by extruding the initial voxels.
    void extrudeVoxelSet();

    // Build an intermediate point cloud with the surface points and normals.
    void buildSurfaceCloud();

    // Build the actual mesh.
    void buildOutputMesh(pcl::PolygonMesh& mesh);

    // Whether a voxel is completely surrounded by others.
    bool isInnerVoxel(const Voxel& v) const;

    // Compute the normal of a given voxel.
    Eigen::Vector3f computeNormal(const Voxel& v);

    // Compute the intersection of the line formed by p1 and p2 with
    // the plane.
    Eigen::Vector3f
    lineIntersectionWithPlane(const Eigen::Vector3f& p1,
                              const Eigen::Vector3f& p2) const;

    // Get the 3-D point that corresponds to a given voxel.
```

```
    Eigen::Vector3f voxelToPoint(const Voxel& v) const;
    // Convert a 3-D point to a voxel.
    Voxel pointToVoxel(const Eigen::Vector3f& p) const;
private:
    PointCloudConstPtr input_cloud_;
    pcl::PointCloud<pcl::PointNormal> surface_cloud_;
    Eigen::Vector4f plane_coeffs_;
    float voxel_size_;
    // Set of filled voxels.
    std::set<Voxel> voxels_;
};
```

The main computation function, shown in Listing 9-6, simply calls each of the intermediate steps introduced before.

Listing 9-6. *Compute a Surface Point Cloud with Normals Out of a Voxel Set*

```
void Extruder::compute(pcl::PolygonMesh& mesh)
{
buildInitialVoxelSet();
    extrudeVoxelSet();
    buildSurfaceCloud();
    buildOutputMesh(mesh);
}
```

The following sections analyze each part in detail.

Construction of a Voxelized Representation

As stated in Chapter 6, it is sometimes convenient to work in the 3-D space as we do in 2-D with images, using small individual elements, called pixels in an image and generally called voxels in a volume. This representation introduces a natural neighborhood between points and allows efficient implementations of searching and filling algorithms.

More concretely, a voxel associated to a 3-D point is a 3-D discrete index. This index is computed according to a predetermined voxel size or resolution. PCL has several classes to deal with voxels, but they are mostly targeting downsampling applications and do not currently offer the level of control that we need here. So we are going to write our own voxel code. Let's start, in Listing 9-7, with the voxel representation itself and the conversions from 3-D points to voxels and vice versa.

Listing 9-7. *Voxel Representation and Conversions from and to 3-DPoints*

```
struct Extruder::Voxel
{
    int c, r, d; // col, row anddepth indices.

    // Constructor with default indices
    Voxel(int c = 0, int r = 0, int d = 0) : c(c), r(r), d(d) {}
```

```
    // Comparison operator, required to store voxels in an std::set
    bool operator<(const Voxel& rhs) const
    {
        if (c < rhs.c) return true;
        if (rhs.c < c) return false;
        if (r < rhs.r) return true;
        if (rhs.r < r) return false;
        if (d < rhs.d) return true;
        if (rhs.d < d) return false;
        return false; // equals, return false.
    }
};

Eigen::Vector3f voxelToPoint(const Voxel& v) const
{
    return Eigen::Vector3f(v.c * voxel_size_,
                           v.r * voxel_size_,
                           v.d * voxel_size_);
}

Voxel pointToVoxel(const Eigen::Vector3f& p) const
{
    return Voxel(p.x() / voxel_size_,
                 p.y() / voxel_size_,
                 p.z() / voxel_size_);
}
```

A voxel is just a set of three integer indices, representing its coordinates in a discrete 3-D space. The conversion to/from floating point 3-D points is simply performed by dividing/multiplying each coordinate by the chosen resolution. There are several alternatives for representing a voxelized point cloud. If the boundaries of the cloud are known in advance and the number of cells is small enough, a fixed-size, dense 3-D matrix may be the most readily adapted. It has a constant time search complexity but has a rather high memory occupation. In our case, since the point cloud is sparse, we chose a representation based on a tree-based set of voxels, which has a $\log(N)$ search complexity but is more flexible and occupies less memory. We rely on the STL here and store the voxels as an std::set<Voxel>.

We can now proceed with the initialization of the voxel set with the object point cloud, as shown in Listing 9-8.

Listing 9-8. *Construction of the Initial Voxel Set*

```
void Extruder :: buildInitialVoxelSet()
{
    for (size_t i = 0; i <input_cloud_->points.size(); ++i)
    {
        Voxel v = pointToVoxel(cloud->points[i].getVector3fMap());
voxels_.insert(v);
    }
}
```

The result of Listing 9-8 can be seen on Figure 9-6, applied on the book object cloud.

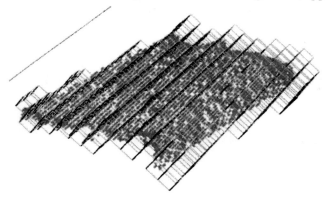

Figure 9-6. *Voxelized point cloud corresponding to the book object. Each voxel is represented by a small cube, and the original point cloud is shown in red. A coarse voxel size of 1cm was chosen here to get a better visualization.*

Filling Unseen Voxels

As shown in Figure 9-6, the initial object point cloud is very sparse, since the Kinect can see only part of the object. The objective of the filling step is to reconstruct the missing parts by extruding the initial voxel set along the plane normal. The algorithm we propose here is pretty simple: for each initial voxel, determine its projection on the table plane, and add all the voxels lying along the line segment, as shown in Figure 9-7. The corresponding code is provided in Listing 9-9.

Listing 9-9. *Extruding the Initial Voxel Set Toward the Table Plane*

```
void Extruder::extrudeVoxelSet()
{
    // Determine the plane normal.
    Eigen::Vector3f plane_normal (plane_coeffs_[0],
                                  plane_coeffs_[1],
                                  plane_coeffs_[2]);
    plane_normal.normalize();

    std::set<Voxel>extruded_voxels;
    // Foreach voxel
    for (std::set<Voxel>::const_iterator it = voxels_.begin();
        it != voxels_.end();
        ++it)
    {
        Eigen::Vector3f origin = voxelToPoint(*it);
        // Projection of the current voxel onto the plane.
```

```
        Eigen::Vector3f end
            = lineIntersectionWithPlane(origin, origin + plane_normal);
        // Walk along the line segment using small steps.
        float step_size = voxel_size_ * 0.1f;

        // Determine the number of steps.
        int nsteps = (end-origin).norm() / step_size;
        for (int k = 0; k < nsteps; ++k)
        {
            Eigen::Vector3f p = origin
                            + plane_normal * (float)step_size * (float)k;
            extruded_voxels.insert(pointToVoxel(p));
        }
    }

    // Set the new set of voxels.
    voxels_ = extruded_voxels;
}
```

This procedure needs to compute the intersection of a line (defined by two points) with a plane, which can be computed using a classical geometrical formula, as shown in Listing 9-10.

Listing 9-10. *Line / Plane Intersection*

```
Eigen::Vector3f

Extruder::lineIntersectionWithPlane(const Eigen::Vector3f& p1,
                                    const Eigen::Vector3f& p2) const

{
    // Plane equation is ax + by + cz + d = 0
    const float a = plane_coeffs_[0];
    const float b = plane_coeffs_[1];
    const float c = plane_coeffs_[2];
    const float d = plane_coeffs_[3];
    float u = a*p1.x() + b*p1.y() + c*p1.z() + d;
    u /= a*(p1.x()-p2.x()) + b*(p1.y()-p2.y()) + c*(p1.z()-p2.z());
    Eigen::Vector3f r = p1 + u * (p2-p1);
    return r;

}
```

Figure 9-7. Left: In the original object cloud, only the front and top parts are seen by the Kinect. Right: This is the point cloud after extrusion along the plane normal. Starting from the original points, the filling algorithm adds all the voxels located between the source points and their intersection with the plane, following the plane normals. The filled voxel centers are shown.

Building a Surface Point Cloud

We now have a compact set of voxels representing the geometry of the object. For many applications, including visualization, having a triangulated mesh of the object surface is more convenient. Before building a triangulated mesh, we first need to extract the voxels that are lying on the object surface, and we will need their normals for the meshing algorithm to work correctly.

Surface voxels can be extracted with a simple neighbor analysis: if the 26 neighbors of a voxel are occupied, it is an inner voxel, and it should be removed. Listing 9-11 shows how to detect an inner voxel.

Listing 9-11. Inner Voxel Detection

```
bool Extruder::isInnerVoxel(const Voxel& v) const
{
    // Walk over the 26-neighborhood
    for (int dc = -1; dc <= 1; ++dc)
    for (int dr = -1; dr <= 1; ++dr)
    for (int dd = -1; dd <= 1; ++dd)
    {
        if (dc == 0 && dr == 0 && dd == 0)
            continue;

        Voxel dv (v.c + dc, v.r + dr, v.d + dd);

        // At least one empty neighbor? Not inner!
        if (voxels_.find(dv) == voxels_.end())
            return false;
    }
    return true;
}
```

Note that if the voxel size is small compared to the point cloud density; some voxels might remain empty within the object volume. A morphological closing on the voxel set can be useful in that case to ensure that the object volume is plain and no inner voxels are missed.

The remaining problem to solve is the computation of the normals of the surface voxels. Again, this can be computed using a simple neighborhood analysis. The idea is the compute the average of the vectors going from a voxel to its empty neighbors voxel center, as illustrated in Figure 9-8. The code is shown in Listing 9-12.

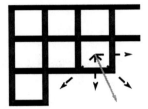

Figure 9-8.*Illustration in 2-D of the computation of normals in a voxelized volume. Vectors going from the voxel center to its empty neighbor centers (dashed) are averaged to get the resulting normal estimation.*

Listing 9-12. *Surface Voxel Normal Computation*

```
Eigen::Vector3f computeNormal(const Voxel& v)
{
    Eigen::Vector3f p = voxelToPoint(v);
    Eigen::Vector3f normal (0, 0, 0);
    int nb_neighbors = 0;
    for (int dc = -1; dc <= 1; ++dc)
    for (int dr = -1; dr <= 1; ++dr)
    for (int dd = -1; dd <= 1; ++dd)
    {
        if (dc == 0 && dr == 0 && dd == 0)
            continue;
        Voxel dv (v.c + dc, v.r + dr, v.d + dd);
        if (voxels_.find(dv) != voxels_.end())
            continue;
        // Empty vector found.
        Eigen::Vector3f neighbor = voxelToPoint(dv);
        Eigen::Vector3f direction = neighbor - p;
        normal += direction;
        nb_neighbors += 1;
    }
    if (nb_neighbors < 1)
        return Eigen::Vector3f(0,0,0);
    normal *= (1.0f / nb_neighbors);
    normal.normalize();
    return normal;
}
```

We can now write the complete surface cloud extraction code, as shown in Listing 9-13.

Listing 9-13. *Compute a Surface Point Cloud with Normals out of a Voxel Set*

```
void Extruder::buildSurfaceCloud ()
{
    for (std::set<Voxel>::const_iterator it = voxels_.begin();
         it != voxels_.end();
         ++it)
    {
        if (isInnerVoxel(*it))
            continue;
        pcl::PointNormal p;
        p.getVector3fMap() = voxelToPoint(*it);
        p.getNormalVector3fMap() = computeNormal(*it);
        surface_cloud_.push_back(p);
    }
}
```

Building a Mesh

Building a triangulated mesh output of the surface point cloud is now straightforward. Since objects are closed shapes, one of the best available algorithms is Poisson reconstruction, based on a global minization of a Poisson equation built from the available points. It is very easy to use, as shown in Listing 9-14.

Listing 9-14. *Compute a Surface Point Cloud with Normals out of a Voxel Set*

```
void Extruder::buildOutputMesh(pcl::PolygonMesh& mesh)
{
    pcl::surface::Poisson<pcl::PointNormal> poisson;
    poisson.setInputCloud(surface_cloud_.makeShared());
    poisson.performReconstruction(mesh);
}
```

The resulting model for the book object is shown in Figure 9-9. Poisson reconstruction significantly smoothes the data, leading to a quite regular mesh. The model is not perfect though; we can observe here the consequence of the imprecision of the Kinect depth maps around object borders, leading to irregular edges.

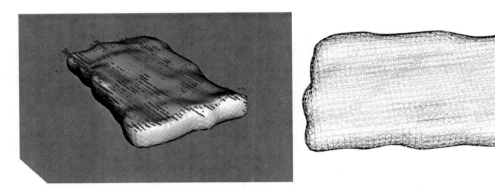

Figure 9-9.*Left reconstructed mesh using Poisson algorithm. Left: Overlaid with the original object point cloud. Right: Wireframe rendering of the estimated model.*

Figure 9-10 shows the result of applying this technique to all the objects of the table scene. The extrusion approximation is pretty good for most of them, with the exception of the ball object because of its spherical shape. And we used only a single Kinect image!

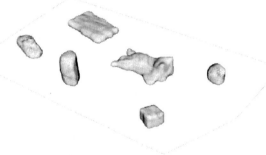

Figure 9-10.*Result of automatic object segmentation and extrusion-based meshing for all the objects of our table sample scene, using a single Kinect depth image*

Acquiring a 3-D Object Model Using Multiple Views

You have seen that a single Kinect frame already gives a lot of information and that approximate models may be built if we have some *a priori* knowledge about the shape of the objects. However, to get accurate models of any kind of object, we need to observe it under different, complementary points of view.

Leveraging multiple Kinect views raises two main issues:

- Estimate the 3-D pose of the camera for each image, in a consistent common coordinate frame.

- Merge the partial point cloud in a single, coherent cloud and/or mesh.

Both of the preceding issues are quite challenging, and we will start with the first one. You already saw a solution for the pose estimation problem in Chapter 8, using SLAM. However, objects are typically much smaller than entire rooms, and it is complicated to get acceptable accuracy of the localization, especially in a scene where the object is predominant. Indeed, objects do not always have enough texture or salient geometric features, thus the tracking can be difficult.

A popular solution for object scanning is to use fiducial markers, and this is the option we are going to work with in the following sections. Fiducial markers are usually seen in the form of black squares with a binary pattern inside and are commonly used in augmented reality applications to robustly track a plane. It is also possible to leverage them to get a cheap object model acquisition system,

---

■ **Note** Another popular solution for object scanning is to use a motorized turntable while keeping the camera fixed. A turntable is usually a rotating plate actuated by a precise motor, so that we know the pose of the object at each frame. Since this requires additional hardware, we will not cover it here.

Overview of a Marker-Based Scanner

The idea is to build a support with some markers printed on it, at fixed positions. We then define a common reference frame on the board that will allow all partial point clouds to be merged in a consistent space. For each viewpoint, the camera pose can be computed by first estimating the 3-D coordinates of the detected marker centers and then apply some classical linear algebra. The corresponding partial point cloud can then be transformed into the common coordinate frame and cropped to the scanning zone. Finally, all the partial point clouds will be merged into a single mesh.

Listing 9-15 provides the interface of the `MarkedViewpoint` class, which aims at generating a partial point cloud transformed into the common reference frame.

Listing 9-15. *The MarkedViewpoint Class*

```
class MarkedViewpoint
{
public:
    MarkedViewpoint()
    {}

    // Compute the aligned viewpoint, with normals, cropped to the
    // scanning zone.
    void computeAlignedPointCloud(
                const cv::Mat3b& rgb_image,
                const cv::Mat1f& depth_image,
                pcl::PointCloud<pcl::PointXYZ>::ConstPtr image_cloud,
                pcl::PointCloud<pcl::PointNormal>& output_cloud);

private:
    // Estimate the 3-D coordinates of the center of a Marker.
    Eigen::Vector3f computeMarkerCenter(const aruco::Marker& marker,
                                const cv::Mat1f& depth_image);
```

```
    // Estimate the homogeneous transform from camera space to
    // the common coordinate frame.
    bool estimateCameraToBoardTransform(Eigen::Affine3f& camera_to_board,
                                        const cv::Mat3b& rgb_image,
                                        const cv::Mat1f& depth_image);

    // Crop the point cloud to the zone above the support.
    void cropCloudToScanningZone (pcl::PointCloud<pcl::PointNormal>& cloud);

    // Helper function to crop a point cloud on the given axis.
    void cropCloudOnAxis(pcl::PointCloud<pcl::PointNormal>& cloud,
                         const char* axis, float min_value, float max_value);

    // Remove points that do not have a valid normal.
    void removePointsWithNanNormals(pcl::PointCloud<pcl::PointNormal>& cloud);

private:
    cv::Size2f board_size_;
    float cx_, cy_, fx_, fy_; // camera intrinsics
};
```

The main procedure is computeAlignedPointCloud, which takes a color image, the corresponding depth image and point cloud, and produces a point cloud in the reference coordinate frame, cropped to the scanning zone. Normals are also computed since this will be useful to apply a meshing algorithm in a future stage. Listing 9-16 provides the code.

Listing 9-16. *The Global Function That Generates an Aligned Point Cloud*

```
void MarkedViewpoint::computeAlignedPointCloud(
    const cv::Mat3b& rgb_image,
    const cv::Mat1f& depth_image,
    pcl::PointCloud<pcl::PointXYZ>::ConstPtr in_cloud,
    pcl::PointCloud<pcl::PointNormal>& out_cloud)
{

    Eigen::Affine3f camera_to_board;
    bool ok = estimateCameraToBoardTransform(camera_to_board,
                                             rgb_image,
                                             depth_image);

    if (!ok)
        return;

    pcl::PointCloud<pcl::Normal> normals;
    computeNormals(in_cloud, normals);
    pcl::concatenateFields(*in_cloud, normals, output_cloud);

    pcl::transformPointCloudWithNormals(out_cloud,
                                        out_cloud,
                                        camera_to_board);
    cropCloudToScanningZone(out_cloud);
}
```

Let us see in details each part.

Building a Support with Markers

We first need to build the physical support with the markers. Several opensource libraries are able to detect fiducial markers. In this chapter, we use Aruco, a lightweight, opensource (BSD license) library that fits perfectly our objectives. It can be downloaded from the project website `http://aruco.sourceforge.net`. The library is able to generate and recognize square markers and outputs the ID and the coordinate of the corners of each detected marker, as shown in Figure 9-11.

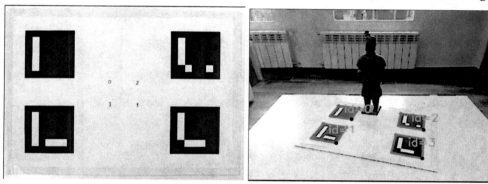

Figure 9-11. Left: Proposed marker board, composed by four markers aligned to form a rectangular shape. They are printed on two different sheets of paper and glued to a carton support. Right: Detected markers using the Aruco library during a scanning session.

As you will see in the next sections, a board with four markers allows a robust estimation of the Kinect pose in each frame, since only three markers have to be successfully detected. For the estimation to be possible, the markers have to be located such that their centers form a rectangular shape. Also, they should be big enough to ensure a good detection. A width of about 10cm is reasonable. The easiest way to build such a support is to print the markers on two sheets of paper and glue them to some support. This leaves enough empty space in the center to put an object without occluding the markers.

■ **Note** Should you move the object or move the camera? To generate new viewpoints, both alternatives are possible. However, we recommend moving the object when possible, because it avoids the risk of having desynchronized depth and color images and motion blur if the camera is moved too fast. It is usually easier to slowly rotate the board support. Also, keeping the camera at the same location ensures a constant illumination, which is desirable if you also want to extract the texture of the objects.

Estimating the 3-DCenter of the Markers in the Camera Space

The first step before estimating the pose of the camera is to extract to 3-D coordinates of the center of each detected markers. Aruco provides us with the coordinates of the markers' corners in the color image, so we can easily deduce the coordinates of the center of each marker in 2-D. To get a 3-D estimate, we need to get the corresponding depth value in the depth image and project the point to 3-D space. In this section, we consider that a depth image aligned with the color image is available. Note that this alignment can be performed automatically in latest libfreenect versions or using the formula introduced in Chapter 2. To make the depth estimation robust to missing and noisy depth values, we compute an average over a 5×5 window around the central pixel, as shown in Listing 9-17.

Listing 9-17. *Estimation of the Center of an Aruco Marker*

```
Eigen::Vector3f computeMarkerCenter(const aruco::Marker& marker,
                                    const cv::Mat1f& depth_image)
{
    cv::Point2f image_center (0,0); // marker center in image coordinates
    for (int i = 0; i < 4; ++i)
        image_center += marker[i]; // marker[i] returns the image coordinates
                                   // of corner i
    image_center /= 4.f; // the center is the mean of the four corners

    float mean_depth = 0;
    int count = 0;
    for (int dy = -3; dy < 3; ++dy) // walk over a 5x5 window
    for (int dx = -3; dx < 3; ++dx)
    {
        int row = roundf(image_center.y + dy);
        int col = roundf(image_center.x + dx);
        if (depth_image(row, col) > 1e-5) // valid depth point
        {
            mean_depth += depth_image(row, col);
            ++count;
        }
    }
    mean_depth /= count;

    return Eigen::Vector3f (mean_depth * (image_center.x - cx_)/fx_,
                            mean_depth * -(image_center.y - cy_)/fy_,
                            -mean_depth);
}
```

This procedure makes use of the calibration parameters estimated in Chapter 2. The marker center in the image is simply the mean of the four corners. We can now proceed with the camera pose estimation itself.

Kinect Pose Estimation from Markers

We first need to define the common reference frame to which all the partial point clouds will be transformed. It can be anywhere on the board, but for convenience reasons, we choose the origin (O_b) to be the center of the bottom left marker, and we align the axes (x_b, y_b, z_b) with the board, as shown in Figure 9-12.

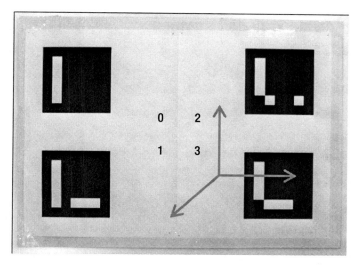

Figure 9-12. This is the common reference frame (x_b, y_b, z_b) used to combine the partial views. All the point clouds extracted at each viewpoint will be transformed into this coordinate system.

Since the relative position of the markers is fixed on the board, it is straightforward to estimate the coordinates (x_{bk}, y_{bk}, z_{bk}) of (x_b, y_b, z_b) in the camera coordinate frame. Let's denote C_0, C_1, C_2 and C_3 the 3-D coordinates of the markers with ID 0, 1, 2 and 3. The following equationshows how the markerboard coordinates can be computed:

$$x_{bk} = \frac{(C_2 - C_0)}{||C_2 - C_0||} = \frac{(C_3 - C_1)}{||C_3 - C_1||}$$
$$y_{bk} = \frac{(C_0 - C_1)}{||C_0 - C_1||} = \frac{(C_2 - C_3)}{||C_2 - C_3||}$$
$$z_{bk} = x_{bk} \wedge y_{bk}$$
$$O_{bk} = C_1 = C_0 - y_{bk} * ||C_2 - C_3||$$

In the preceding example, $\wedge$ is the cross-product operator. Since we have at least two different ways of computing each axis coordinates, we need only three detected markers to get an estimate. It is now straightforward to compute the transformation $H_{b \to k}$ that transforms a point from the common reference frame to the camera frame, as detailed in the following equation:

$$H_{b \to k} = \begin{bmatrix} x_{bk}.x & y_{bk}.x & z_{bk}.x & O_{bk}.x \\ x_{bk}.y & y_{bk}.y & z_{bk}.y & O_{bk}.y \\ x_{bk}.z & y_{bk}.z & z_{bk}.z & O_{bk}.z \\ 0 & 0 & 0 & 1 \end{bmatrix}$$

The final transform from camera space to the common frame is then $H_{k \to b} = [H_{b \to k}]^{-1}$. This leads us to Listing 9-18.

Listing 9-18. Camera Pose Estimation Based on Fiducial Marker Detection

```
bool MarkedViewpoint::estimateCameraToBoardTransform(
    Eigen::Affine3f& camera_to_board,
    const cv::Mat3b& rgb_image,
    const cv::Mat1f& depth_image)
{
    aruco::MarkerDetector detector;
    std::vector<aruco::Marker> markers;
    detector.detect(rgb_image, markers); // get the list of markers from Aruco
    if (markers.size() < 3)
        return false; // at least 3 markers needed.

    // Index markers by id name.
    // markers_by_id[k] corresponds to the marker with id k, or 0
    // if not detected.
    std::vector<aruco::Marker*> markers_by_id (4, (aruco::Marker*) 0);
    for (size_t i = 0; i < markers.size(); ++i)
        markers_by_id[markers[i].id] = &markers[i];

        // Get the 3-D coordinates of the marker centers.
        std::vector<Eigen::Vector3f,
                    Eigen::aligned_allocator<Eigen::Vector3f> >
                    marker_centers(4);
        for (size_t i = 0; i < markers_by_id.size(); ++i)
        {
            if (markers_by_id[i])
                marker_centers[i] = computeMarkerCenter(*markers_by_id[i],
                                                        depth_image);
        }

        Eigen::Vector3f x_axis; // x_{bk}
        Eigen::Vector3f y_axis; // y_{bk}
        Eigen::Vector3f origin;// O_{bk}

        if (markers_by_id[0] && markers_by_id[2])
            x_axis = marker_centers[2] - marker_centers[0];
        else
            x_axis = marker_centers[3] - marker_centers[1];

        if (markers_by_id[0] && markers_by_id[1])
            y_axis = marker_centers[0] - marker_centers[1];
        else
            y_axis = marker_centers[2] - marker_centers[3];
```

```
    if (markers_by_id[1])
        origin = marker_centers[1];
    else
        origin = marker_centers[0] - y_axis;

    // Store the board size, will be useful for cropping.
    board_size_.width = x_axis.norm();
    board_size_.height = y_axis.norm();

    x_axis.normalize();
    y_axis.normalize();
    Eigen::Vector3f z_axis = x_axis.cross(y_axis);

    // build H_{b→k}
    Eigen::Affine3f board_to_camera (Eigen::Affine3f::Identity());
    board_to_camera.matrix().block<3,1>(0,0) = x_axis;
    board_to_camera.matrix().block<3,1>(0,1) = y_axis;
    board_to_camera.matrix().block<3,1>(0,2) = z_axis;
    board_to_camera.translation() = origin;
    camera_to_board = board_to_camera.inverse(); // return H_{k→b}

    return true;
}
```

Cleaning and Cropping the Partial Views

Once a point cloud has been transformed into the common frame, it is very easy to crop it to the scanning region. The limits on the X axis are zero and the board width, on the Y, zero and the board height, and on the Z axis, we put them at (0.03, 0.4) so that points lying at less than 3 cm from the board are not included; neither are the points lying higher than 40cm. This way, the board itself will not be part of the final point cloud. To make sure that you do not lose part of the objects while doing so, we recommend placing them on a small transparent pedestal.

The cropping code is easy to implement using the PassThrough filter from PCL, as shown in Listing 9-19.

Listing 9-19. Camera Pose Estimation Based on Fiducial Marker Detection

```
void MarkedViewpoint::cropCloudToScanningZone(
    pcl::PointCloud<pcl::PointNormal>& cloud)
{
    cropCloudOnAxis(output_cloud, "x", 0.00, board_size_.width);
    cropCloudOnAxis(output_cloud, "y", 0.00, board_size_.height);
    cropCloudOnAxis(output_cloud, "z", 0.03, 0.4);
}

void MarkedViewpoint::cropCloudOnAxis(
    pcl::PointCloud<pcl::PointNormal>& cloud,
```

```
        const char* axis, float min_value, float max_value)
{
    pcl::PassThrough<pcl::PointNormal> bbox_filter;
    bbox_filter.setFilterFieldName(axis);
    bbox_filter.setFilterLimits(min_value, max_value);
    bbox_filter.setInputCloud(cloud.makeShared());
    bbox_filter.filter(*cloud);
}
```

The result of the cropping process is shown in Figure 9-13. By using a transparent pedestal, we make sure that only points belonging to the object to be scanned are actually kept.

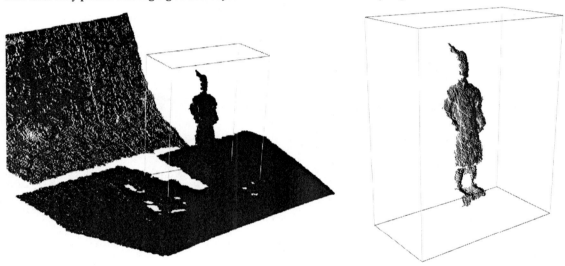

Figure 9-13. *Cropping process. Left: Raw point cloud with the cropping box represented with a wireframe cube, which is aligned with the markers. Right: Points left after cropping.*

Merging the Point Clouds

By using the `MarkedViewpoint` class on a set of different viewpoints, we end up with a collection of point clouds, aligned in a common coordinate frame. Figure 9-14 gives an example of output for the statue object. As can be observed, the partial point clouds do not correspond exactly because of the accumulation of Kinect depth estimations and the inherent imprecision of the position of the markers in the image.

To improve the quality of the reconstruction, a global refinement step is necessary. This is typically done by iteratively registering the views between themselvesand defining some kind of global metric to be minimized. Given that there is no such algorithm available in PCL yet and that implementing our own multiview registration is out of the scope of this book, we rely on Meshlab here. Meshlab is an opensource visualization and mesh manipulation software developed at the Visual Computing Lab (ISTI-CNR). It can be downloaded from http://meshlab.sourceforget.net, and it implements a wide range of algorithms, among which is a global alignment tool that we will use here.

We consider here that you could generate a set of `.ply` files (e.g., using `pcl::io::PlyWriter`), one per viewpoint, using the `MarkedViewpoint` class. These can be loaded into Meshlab by selecting File and then Import Mesh. Next, select View Show Layer Dialogto open a small window that allows to hide or show the point clouds individually. You should see something similar to Figure 9-14.

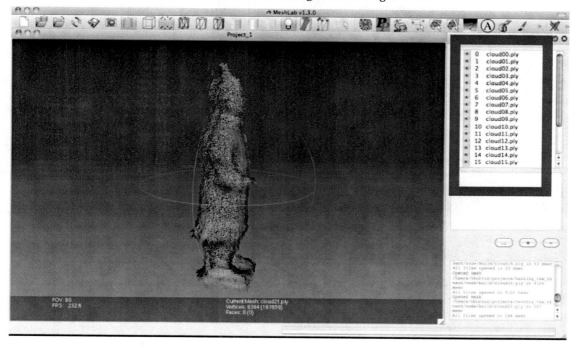

Figure 9-14.*Layer stack in Meshlab*

Now, we need to open the alignment tool, using the yellow A button (see Figure 9-15). Meshlab gives us the opportunity to manually initialize the position of the clouds to help the alignment algorithm to converge. Since we got a pretty good initialization with the marker-based pose estimation, we can validate the initial positions directly by clicking "Glue here all Meshes".We can now proceed with the registration itself by clicking Process. Since the algorithm is iterative, you might need to launch the process several times until it converges.

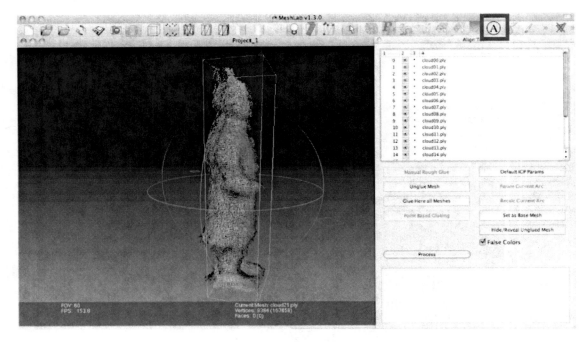

Figure 9-15.*Align tool in Meshlab*

Figure 9-16 shows the final point cloud after alignment. We can now flatten all the layers into a single one using Filters   Layer and Attribute Management   Flatten Visible Layers, without forgetting to check the "Keep unreferenced vertices" checkbox. A triangulated mesh can then be built using the Poisson surface reconstruction algorithm, also implemented in Meshlab under Filters Point Sets Surface reconstruction: Poisson. Figure 9-16 (right) gives the final output for the statue object.

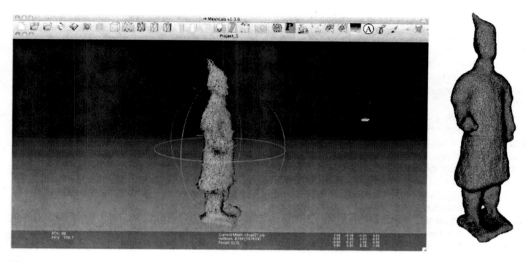

Figure 9-16.Final point cloud in Meshlab (left) and result of the Poisson reconstruction (right)

Getting a Better Resolution

Using a set of viewpoints, we could get a better model than using a single image. But the level of detail is still quite low because of several factors:

- The Kinect accuracy is about 2 or 3 millimeters at short range. For small objects, this can be significant and prevents the reconstruction of small details.

- The accuracy of the marker-based pose estimation depends on the resolution of the color image. Using the standard 640×480 RGB mode, an error of 1 pixel can lead to a bias of more than 1 millimeter in the marker corner position. Cumulated with the Kinect error in the depth estimation, the center of the marker can have a bias of several millimeters.

There are several options to overcome the preceding limitations and build high-resolution models with Kinect.The first, straightforward improvement is to use the Kinect in high-resolution mode, which outputs RGB images of 1280 × 1024 pixels (but at only 10 frames/second). This allows a more precise estimation of the marker pose.

Regarding the depth estimation itself, the precision can be improved by taking images from many slightly different viewpoints. The expected coherence between the consecutive frames can lead to a substantial increase of the precision. This approachis sometimes referredto as "super resolution techniques," and the fusion of many images is also at the heart of the very impressive Microsoft Kinect Fusion project.

Detecting Acquired Objects

You have seen several methods for acquiring a model of a real-world object. We now consider the problem of detecting the known objects in a new scene. This task is challenging for several reasons. First, the new viewpoint will be different, resulting in a different appearance and shape of the object in the image. The lighting conditions might also have changed, and there might be some occlusions due to other elements of the scene. Finally, background clutter can lead to false detection.

There are two main approaches for object recognition and localization. One approach is to presegment the scene to determine the location of the candidate objects and then use some global appearance comparison with the reference model. If the similarity is high enough, the object is recognized. The precise estimation of the pose of the object is then estimated by trying to align the model with the image.

This approach is very difficult in practice using a regular color camera, because the initial presegmentation is very challenging in the presence of background clutter. This leads to the development of an alternative set of methods based on local feature matching, as you saw in Chapter 8 for the SLAM application. A robust interest points detector and powerful local descriptors have been proposed in the last decade, leading to satisfying results when the objects are textured enough.

When using a Kinect, the problem takes a slightly different perspective. Methods based on local descriptors can still be used thanks to the color camera, but methods based on global appearance become more practical thanks to the depth image, which enables an easier presegmentation by taking into account the occlusions. To focus on the new information provided by the Kinect, we propose in this chapter to detect and estimate the pose of the objects we previously acquired using a global descriptor based on geometric features only. The color channel will not be used.

Detection Using Global Descriptors

As said previously, when using global descriptors, an initial segmentation is generally required. The objective is then to recognize if the segmented object correspond to an object of the database, and if it does, estimate its pose. A sample scene and expected output are presented in Figure 9-17.

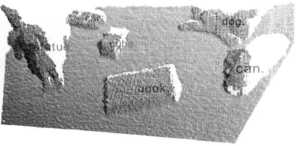

Figure 9-17.Recognition and pose estimation of objects lying on a table. The model point clouds, extracted from a single view are overlaid on the recognized clusters, using the estimated pose.

To get an initial segmentation of the scene into clustered point clouds, we use the `TableTopDetector` class introduced earlier in this chapter. The general principle is to compute a descriptor for each object cloud and compare it with the set of descriptors computed on the previously acquired models. There are two families of global descriptor. The first family tries to find a representation that is viewindependent, thereby allowing a match of the model in the candidate image under any pose. This comes at the price of a much lower discriminant power, and thus the current trend is to propose viewpoint-dependent descriptors, computed for a set of representative viewpoints of the model. The database of descriptors is then much larger, since each model leads to a high number of descriptors, typically about a hundred.

This approach is still preferred for most applications, since there are data structures and algorithms (e.g.,Kd-trees) that allow finding the closest descriptor in a big database very efficiently.

To illustrate the main principles of this second category of techniques, we show here an example of point cloud recognizer. For the sake of simplicity, we only consider a single view for each object, and the model only consists of the extracted object clouds in the TableTopDetector. Then, the tabletop detector is run again on new scenes, and the objective is to find for each new object cloud the closest model in the database and estimate the pose change. We rely here on the Viewpoint Feature Histogram (VFH) descriptor implemented in PCL, which is a robust viewpoint-dependent descriptor.

Listing 9-20 provides the interface of the CloudRecognizer class.

Listing 9-20. *Interface of the CloudRecognizer Class*

```
class CloudRecognizer
{
public:
    // Data associated to a model in the database.
    struct Model
    {
        // Point cloud of the model view.
        pcl::PointCloud<pcl::PointNormal>::ConstPtr cloud;

        // Corresponding VFH descriptor, a vector of 308 float values.
        pcl::VFHSignature308descriptor;
        // Name of the model.
        std::string name;
    };

public:
    // Insert a new model in the database.
    void addModel(pcl::PointCloud<pcl::PointNormal>::ConstPtr cloud,
                const std::string& name)

    // Return the model that is more similar to the given cloud, and its pose.
    const Model* recognizeCloud(
        pcl::PointCloud<pcl::PointNormal>::ConstPtr cloud,
        Eigen::Affine3f&model_pose);

private:
    // Compute the pose of the recognized model.
    Eigen::Affine3f
    computeModelPose(pcl::PointCloud<pcl::PointNormal>::ConstPtr cloud,
                    const Model* model) const;

    // Find the model that is most similar to the given cloud descriptor.
    const Model*
    findClosestModel(const pcl::VFHSignature308& descriptor) const;

    // Compute the distance between two descriptors.
    float computeDistance(const pcl::VFHSignature308& d1,
```

```
                        const pcl::VFHSignature308& d2) const;

    // Compute the VFH descriptor of the given cloud.
    void computeDescriptor(pcl::PointCloud<pcl::PointNormal>::ConstPtr cloud,
                           pcl::VFHSignature308& descriptor) const;

private:
    // Database of models.
    std::vector<Model> models_;
};
```

This class has two public methods: one to add a new model to the database and one to get the model that is most similar to a point cloud. The latter function assumes that the cloud to be recognized has already been segmented and corresponds to a candidate object. Let's first see how model are added into the database, as shown in Listing 9-21.

Listing 9-21. *Adding a Model to the Database*

```
void CloudRecognizer::addModel(
    pcl::PointCloud<pcl::PointNormal>::ConstPtr cloud,
const std::string& name)
{
    Model model;
    model.cloud = cloud;
    model.name = name;
    computeDescriptor(model.cloud, model.descriptor);
    models_.push_back(model); // add the new model to the database
}
```

The method simply fills the field of a new Model object and adds it to the database. The recognizeCloudmethod is almost as simple, as shown in Listing 9-22.

Listing 9-22. *Adding a Model to the Database*

```
const Model* CloudRecognizer::recognizeCloud(
    pcl::PointCloud<pcl::PointNormal>::ConstPtr cloud,
    Eigen::Affine3f& model_pose)
{
    pcl::VFHSignature308 cloud_descriptor;
    computeDescriptor(cloud, cloud_descriptor);
    const Model* closest_model = findClosestModel(cloud_descriptor);
    assert(closest_model != 0); // should be a valid model.
    model_pose = computeModelPose(cloud, closest_model);
    return closest_model;
}
```

The recognizeCloud method computes the descriptor of the candidate cloud, retrieves the most similar model in the database, and finally, computes the pose of the model. The actual recognition code is found in the computeDescriptor and findClosestModel methods, as shown in Listing 9-23.

Listing 9-23. *Computing the VFH Descriptor of a Point Cloud and Comparing It to Others*

```
void CloudRecognizer::computeDescriptor(
    pcl::PointCloud<pcl::PointNormal>::ConstPtr cloud,
    pcl::VFHSignature308& descriptor) const
{
    // Structure used for efficient nearest neighbor search.
    pcl::KdTree<pcl::PointNormal>::Ptr search_tree
        (new pcl::KdTreeFLANN<pcl::PointNormal>);

    // holder for the descriptors
    pcl::PointCloud<pcl::VFHSignature308> descriptors;

    pcl::VFHEstimation<pcl::PointNormal,
                       pcl::PointNormal,
                       pcl::VFHSignature308> extractor;
    extractor.setInputCloud(subsampled_object_cloud);
    extractor.setInputNormals(object_normals);
    extractor.setSearchMethod(search_tree);
    extractor.compute(descriptors);
    assert(descriptors.points.size() == 1); // should be single descriptor
    descriptor = descriptors.points[0];
}

const Model* CloudRecognizer::findClosestModel(
    const pcl::VFHSignature308& descriptor) const
{
    const Model* best_model = 0;
    float best_distance = FLT_MAX;

    // Brute force comparison with all the descriptors in the database.
    for (size_t i = 0; i < models_.size(); ++i)
    {
        float dist = computeDistance(models_[i].descriptor, descriptor);
        if (dist < best_distance)
        {
            best_distance = dist;
            best_model = &models_[i];
        }
    }
    return best_model;
}

// Compute the L2 distance between two descriptors.
float CloudRecognizer::computeDistance(
    const pcl::VFHSignature308& d1,
    const pcl::VFHSignature308& d2) const
{
```

```
    float dist = 0;
    for (int i = 0; i <308; ++i)
    {
        float diff = d1.histogram[i] - d2.histogram[i];
        dist += diff * diff;
    }
    return dist;
}
```

The VFHEstimation class from PCL extracts the descriptor of a given point cloud. It requires a point cloud with normals and returns a point cloud of descriptors. Since VFH is global, only one descriptor should be returned. To find the closest descriptor in the database, a simple brute force comparison is performed, using a Euclidian distance between the descriptors.

Note that if some candidate clouds don't correspond to any model, the minimal distance could be threshold to avoid false detections.

Estimating the Pose of a Recognized Model

Once a model has been recognized, most applications also need to estimate its pose. Since we used a viewpoint-dependent descriptor, we know that the object point cloud has a similar viewpoint as the model. This hypothesis makes it possible to use the Iterative Closest Point (ICP) family of algorithms to align the model with the observed point cloud. ICP and its variants are probably the most popular approaches to register two 3-D point clouds when a good initialization can be obtained. The idea of the algorithm is to iteratively:

1.   Associate the points of the first cloud with the points of the second one, generally by picking the closest one.

2.   Find the transformation that minimizes the distance between the associated points.

3.   Repeat until convergence.

The algorithm is illustrated in Figure 9-18. If we run it directly on the model and observed point cloud, it is unlikely to converge, because the point cloud may have a very big initial distance.

Figure 9-18. *Illustration of four iterations of the ICP algorithm in 2-D to align two points clouds with a rigid transformation. At each iteration, the nearest neighbors of the source cloud points are looked for in the target cloud. Then, the transformation is optimized to best align these correspondences. This leads to a new set of nearest neighbords in the next iterations, progressively leading to the right alignment.*

We first align their centroids with a simple translation and then run ICP. The algorithm is implemented in PCL, so there is little code to write here, as shown in Listing 9-24.

Listing 9-24. *Aligning the Model Cloud with the Observed Object Cloud*

```
Eigen::Affine3f computeModelPose(
    pcl::PointCloud<pcl::PointNormal>::ConstPtr observed_cloud,
    const Model* model) const
{
    // Compute the centroids of both clouds
    Eigen::Vector4f observed_centroid;
    pcl::compute3DCentroid(*observed_cloud, observed_centroid);

    Eigen::Vector4f model_centroid;
    pcl::compute3DCentroid(*model->cloud, model_centroid);

    // Transformation required to align the centroids
    Eigen::Affine3f transform (Eigen::Affine3d::Identity());
    transform.translate(observed_centroid- model_centroid);

    // Transform the model point cloud
    pcl::PointCloud<pcl::PointNormal> centered_model_cloud;
    pcl::transformPointCloudWithNormals(*model->cloud,
                                        centered_model_cloud,
                                        transform);

    // Apply the ICP algorithm to refine the alignment.
    pcl::IterativeClosestPoint<pcl::PointNormal, pcl::PointNormal> reg;
    reg.setMaximumIterations (50);
    reg.setMaxCorrespondenceDistance (0.05f);
    reg.setInputCloud (centered_model_cloud.makeShared());
    reg.setInputTarget (observed_cloud);
    pcl::PointCloud<pcl::PointNormal> aligned_cloud;
    reg.align (aligned_cloud);

    // Compute the final transformation: centroid alignment + ICP
    transform = reg.getFinalTransformation() * transform.matrix();
    return transform;
}
```

The `pcl::IterativeClosestPoint` class has several parameters that might need some tuning to get satisfying results. The maximum number of iterations should be big enough to ensure the convergence. The transformation epsilon parameter is used to determine whether the algorithm has already converged. The maximum correspondence distance parameter is the most important one. It sets the distance above which two points cannot be associated in the first step on the algorithm. This enables to align clouds that do not fully overlap, by discarding points that are to far away from the other cloud. The threshold should be high enough to ensure a convergence when the initialization is not very accurate, but low enough to avoid outliers to lead to a wrong registration.

Summary

The Kinect opens new possibilities regarding the fast acquisition of 3-D models of objects. Using a single Kinect image only, it is already possible to reconstruct a good approximate model of an object lying on a table. If more precision is required, multiple Kinect views can be merged using fiducials markers to estimate the pose of the camera in each image, and a complete mesh can be reconstructed after proper alignment. The 3-D models acquired by Kinect can be further recognized in new scenes thanks to powerful 3-D descriptors that allow you to robustly recognize point clouds, even with small viewpoint changes.

Multiple Kinects

In this chapter, we will discuss in depth why you would want to use multiple Kinects, the issues with running multiple Kinects at the same time, and some of the solutions to those problems, including the following:

- The weaknesses of the sensor and how to overcome them with multiple Kinects

- How to stop interference from destroying your accuracy

- How to make an autocalibrating multiple Kinect system

Why Multiple Kinects?

Given all of the amazing things you can do with a single Kinect, it may be hard to believe that you'd want to use more than one. However, there are some key limitations in the sensor that you may want to overcome or some expansions you may want to undergo to fit your application. We'll discuss a few of these limitations and possibilities and talk about how multiple devices can help.

The Kinect Has a Limited Field of View

The Kinect's field of view is limited, as is its depth reach. The field of view of the Kinect's RGB camera is close to 62.7 degrees, while the IR camera has a field of view of 57.8 degrees. At a maximal range of 10 feet, the depth image is still only 14.45 feet wide, which is barely enough to cover one side of a medium-sized room at distance. As you get closer to the sensor, this narrows significantly. At the average working distance of 7 feet, you only have 10.11 feet. By adding additional Kinects, you can widen your field of view. You just chain them along as necessary.

The Kinect Fills Data from a Single Direction Only

The 3D data generated from the Kinect comes from a single direction—a point in space oriented around the center of the Kinect head. The normals of the returns are necessarily strongest in that specific direction. This results in better returns the more perpendicular the imaged object is to the sensor's normal (or how closely parallel their respective normals are). This also results in one-sided objects or people. You have to move the Kinect in order to generate the opposite side. With multiple Kinects, you can add the data captured together, producing a more complete scene. You can also eliminate the mixed pixel effect by then culling said data using voxelization techniques.

The Kinect Casts Depth Shadows in Occlusions

A corollary to the single direction filling is the depth shadow cast by occlusions in the scene by one object over another or even just the background. Depending on the distance of the object from the sensor and the final backdrop, this shadow can be enormous, making much of the scene invisible to the Kinect. This is even worse for applications that work on probability because the areas that are occluded must be marked as being unknown, a state that provides no information to the overall calculation of the state of the system. See Figure 10-1 for an example of an occlusion caused by an object across a table. With every extra Kinect, you can fill the occluded areas, bringing structured light to them and then combining the captures.

Figure 10-1. An occlusion caused by an object creates a depth shadow on a table.

What Are the Issues with Multiple Kinects?

Given all of these problems and the possibility to solve them with multiple Kinects, why not just hook up multiple Kinects and get them all running every time? There are several issues standing in the way of implementing a multiple Kinect system. However, while these obstacles do exist, there are ways to mitigate or even eliminate them as problems. Let's take a quick look at what these issues are and then explore ways to handle them.

Hardware Requirements

Each Kinect requires its own USB hub. While most computers have two, thus enabling you to use two Kinects at a time, more than that will require extra PCI bus USB cards. Also, the more Kinects you add, the more data you'll get, which will rapidly consume system resources. Each Kinect produces 30 Mb/sec of data, but when you have to handle each frame through PCL, this quickly adds up. In order to handle this greater flow of data, advanced techniques like using the GPU in your graphics card are necessary to maintain a steady framerate. This is really on the cutting edge and isn't for normal users. If you'd like to know more, head over to http://docs.pointclouds.org/trunk/namespacepcl_1_1gpu.html.

Interference Between Kinects

When multiple Kinects illuminate the same area with their IR projectors, interference is created. The random pattern that each Kinect produces from its diffraction grating is burned into the hardware. When extra dots are added, noise is added into this pattern, and it confuses the depth differencing algorithm. This results in NaN values (bad data) for the depth or small blotches of depth scattered randomly over the entire depth window.

Calibration Between Kinects

Using multiple Kinects requires converting each camera's base frame (or camera frame) into the world frame (a frame of reference that the "whole world" or your scene shares). This conversion requires knowing precisely the distance between the centers of the Kinect cameras or the position of both Kinects in the world frame. There are a few ways to accomplish this calibration, as discussed in Chapter 3, but the calibration gets more complicated with multiple systems.

Interference

Interference due to multiple Kinects is caused by the mixing of two IR illumination patterns. See Figure 10-2 for a regular IR pattern and Figure 10-3 for an image of mixed illumination patterns. Do you see how the regularity in the mixed figure is disrupted? This is called noise. Interestingly, in both Kinects, the noise is the other Kinect's pattern. This noise confuses the depth differencing algorithm in the Kinect. Instead of detecting just the disruption in its own field, the Kinect now has a set of points that should not exist where they exist. This causes all sorts of effects, but the two primary effects are shown in Figure 10-4 and Figure 10-5. Figure 10-4 shows "holes" in the detected depth field and Figure 10-5 shows "splotches" generated in areas of no depth detection. The holes can be smoothed over with a variety of methods, but the splotches are far more concerning. They rapidly create a blizzard of false positives that can overwhelm your system. A size filter may help, eliminating any set of voxels or points that don't have a connected size greater than X, but this adds to your computation time.

Figure 10-2. *Single Kinect IR pattern*

Figure 10-3. *Dual Kinect IR pattern*

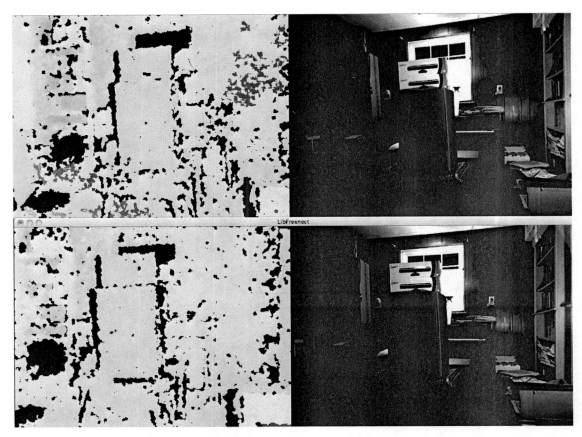

Figure 10-4. Interference holes in the depth return (also note the scattered red—that's also bad data)

Figure 10-5. Interference splotches in scene

The interference in a scene is not necessarily always crippling. Depending on the angle and distance between two Kinects, you can have very good results or amazingly bad ones. What follows in Table 10-1 is a series of images of a scene captured from two different Kinects at a set of angles and distances from each other. The distances are from sensor to sensor while the angles are negative, pointing further towards the center of the scene. Figure 10-6 is an example of a scene without interference for comparison.

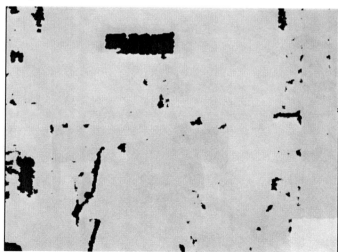

Figure 10-6. Scene without interference

Table 10-1. *Angles/Distances*

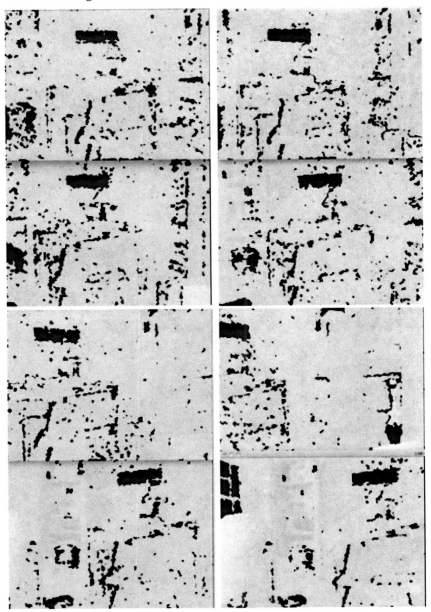

28 cm between image centers from left to right, top to bottom 0, 15, 30, 45 degrees

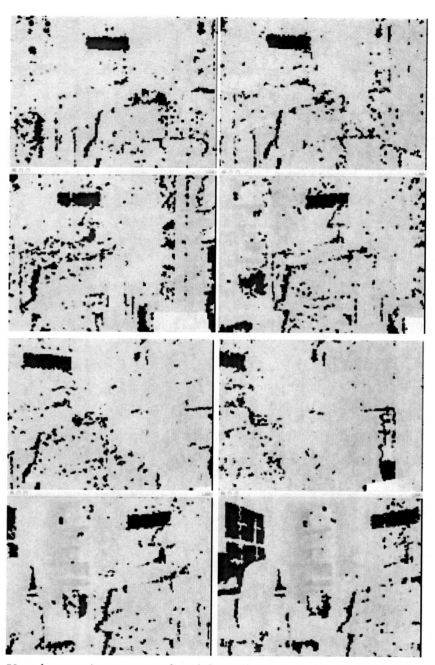

50 cm between image centers from left to right, top to bottom 0, 15, 30, 45 degrees

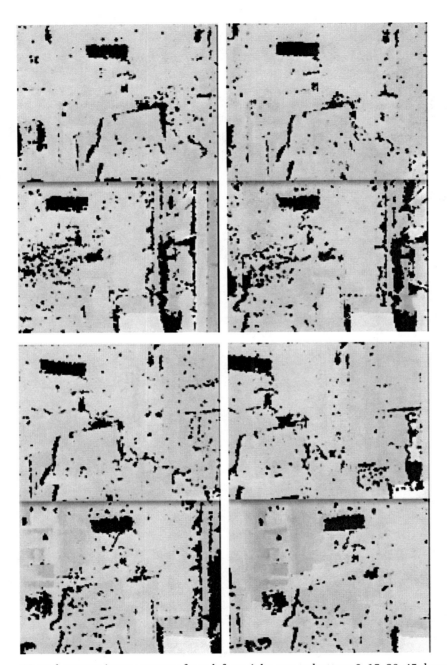

75 cm between image centers from left to right, top to bottom 0, 15, 30, 45 degrees

PROJECT: ELIMINATE INTERFERENCE BETWEEN TWO KINECTS

As you can see, sometimes interference between two Kinects is rather awful. We decided that we needed a way to eliminate interference from the images.

We had two possible methods to achieve this. One was to turn off or diminish the illuminator directly via electronics. The other was to build a set of mechanical shutters to interrupt the light from reaching the IR camera.

The first idea suffers from trying to interface with the laser diode. Laser diodes are both finicky and dangerous. The diode itself is very tightly controlled to always output the same wavelength via both power and cooling. Also, opening the case could potentially expose us to high-powered IR laser light. The laser inside the Kinect is NOT eye safe without the diffraction grating.

So we went with our second idea of building a set of mechanical shutters. We needed to quickly prototype this design, so we chose Legos (Technic Legos, but still Legos). First, we did some quick testing with a box fan, as shown in Figure 10-7. Did the interruption of the IR system improve quality? The answer was yes. However, it revealed some concerns and problems.

Figure 10-7. Box fan test

We need to interrupt the IR illuminator, not the camera. This quickly became clear when the interrupted camera (but exposed illuminators) still added noise. We also realized that timing was a serious issue. There are three time bases in the system: the shutters and each of the Kinects. There is no way to coordinate the two Kinect's time bases, so there is no reason to try to coordinate the shutters.

After giving this system even more though, we had a breakthrough. We realized that only the phase of the shutters matter. Therefore, we needed a way to detect the phase. The IR camera was a perfect solution. By interrupting both the IR camera and IR illuminator on a single Kinect at the same time, we can decide whether or not to include that image in that frame's 3D output. Testing with index cards revealed that we can inform the region of valid depth by looking at the blocked-out data.

So we built the hardware shutter system shown in Figures 10-8, 10-9, and 10-10. The motor on the right drives the shaft that turns the two rotary-to-linear motion systems. These raise and lower the two pairs of shutters in a synchronized fashion, blocking and uncovering the IR camera and IR illuminator pair for each Kinect in turn.

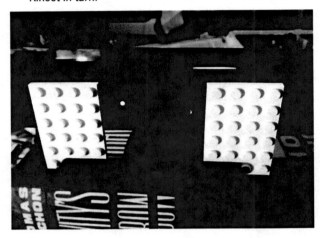

Figure 10-8. Shutters closed from the front

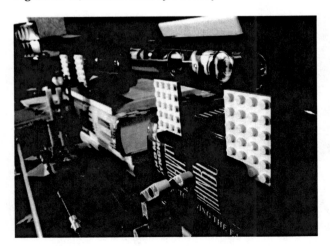

Figure 10-9. Shutters open

Figure 10-10. Shutters closed

The last step was building the code to handle the synchronization. We only update the cloud for a Kinect when the OTHER Kinect's IR camera returns 90% zeros or better. See the code in Listing 10-1 for details.

Listing 10-1. OKShutter.cpp

```
/*
 * This file is part of the OpenKinect Project. http://www.openkinect.org
 *
 * Copyright (c) 2010 individual OpenKinect contributors. See the CONTRIB file
 * for details.
 *
 * This code is licensed to you under the terms of the Apache License, version
 * 2.0, or, at your option, the terms of the GNU General Public License,
 * version 2.0. See the APACHE20 and GPL2 files for the text of the licenses,
 * or the following URLs:
 * http://www.apache.org/licenses/LICENSE-2.0
 * http://www.gnu.org/licenses/gpl-2.0.txt
 *
 * If you redistribute this file in source form, modified or unmodified, you
 * may:
 *   1) Leave this header intact and distribute it under the same terms,
 *      accompanying it with the APACHE20 and GPL20 files, or
 *   2) Delete the Apache 2.0 clause and accompany it with the GPL2 file, or
 *   3) Delete the GPL v2 clause and accompany it with the APACHE20 file
 * In all cases you must keep the copyright notice intact and include a copy
 * of the CONTRIB file.
 *
 * Binary distributions must follow the binary distribution requirements of
 * either License.
 */
```

```cpp
#include <iostream>
#include <libfreenect.hpp>
#include <pthread.h>
#include <stdio.h>
#include <string.h>
#include <cmath>
#include <vector>
#include <ctime>
#include <boost/thread/thread.hpp>
#include "pcl/common/common_headers.h"
#include "pcl/common/eigen.h"
#include "pcl/common/transforms.h"
#include "pcl/features/normal_3d.h"
#include "pcl/io/pcd_io.h"
#include "pcl/visualization/pcl_visualizer.h"
#include "pcl/console/parse.h"
#include "pcl/point_types.h"
#include <pcl/kdtree/kdtree_flann.h>
#include <pcl/surface/mls.h>
#include "boost/lexical_cast.hpp"
#include "pcl/filters/voxel_grid.h"
#include "pcl/octree/octree.h"
//OPENCV Includes
#include "opencv2/core/core.hpp"
#include "opencv2/core/eigen.hpp"

using namespace cv;
using namespace std;

///Mutex Class
class Mutex {
public:
    Mutex() {
            pthread_mutex_init( &m_mutex, NULL );
    }
    void lock() {
            pthread_mutex_lock( &m_mutex );
    }
    void unlock() {
            pthread_mutex_unlock( &m_mutex );
    }

    class ScopedLock
    {
            Mutex & _mutex;
    public:
            ScopedLock(Mutex & mutex): _mutex(mutex)
            {
                    _mutex.lock();
            }
            ~ScopedLock()
            {
```

```
                        _mutex.unlock();
                }
        };
private:
        pthread_mutex_t m_mutex;
};


///Kinect Hardware Connection Class
/* thanks to Yoda---- from IRC */
class MyFreenectDevice : public Freenect::FreenectDevice {
public:
        MyFreenectDevice(freenect_context *_ctx, int _index)
                 : Freenect::FreenectDevice(_ctx, _index),
depth(freenect_find_depth_mode(FREENECT_RESOLUTION_MEDIUM,
FREENECT_DEPTH_REGISTERED).bytes),m_buffer_video(freenect_find_video_mode(FREENECT_RESOL
UTION_MEDIUM, FREENECT_VIDEO_RGB).bytes), m_new_rgb_frame(false),
m_new_depth_frame(false)
        {

        }
        //~MyFreenectDevice(){}
        // Do not call directly even in child
        void VideoCallback(void* _rgb, uint32_t timestamp) {
                Mutex::ScopedLock lock(m_rgb_mutex);
                uint8_t* rgb = static_cast<uint8_t*>(_rgb);
                std::copy(rgb, rgb+getVideoBufferSize(), m_buffer_video.begin());
                m_new_rgb_frame = true;
        };
        // Do not call directly even in child
        void DepthCallback(void* _depth, uint32_t timestamp) {
                Mutex::ScopedLock lock(m_depth_mutex);
                depth.clear();
                uint16_t* call_depth = static_cast<uint16_t*>(_depth);
                for (size_t i = 0; i < 640*480 ; i++) {
                        depth.push_back(call_depth[i]);
                }
                m_new_depth_frame = true;
        }
        bool getRGB(std::vector<uint8_t> &buffer) {
                Mutex::ScopedLock lock(m_rgb_mutex);
                if (!m_new_rgb_frame)
                        return false;
                buffer.swap(m_buffer_video);
                m_new_rgb_frame = false;
                return true;
        }

        bool getDepth(std::vector<uint16_t> &buffer) {
                Mutex::ScopedLock lock(m_depth_mutex);
                if (!m_new_depth_frame)
                        return false;
```

```
                buffer.swap(depth);
                m_new_depth_frame = false;
                return true;
        }

    private:
        std::vector<uint16_t> depth;
        std::vector<uint8_t> m_buffer_video;
        Mutex m_rgb_mutex;
        Mutex m_depth_mutex;
        bool m_new_rgb_frame;
        bool m_new_depth_frame;
};


///Start the PCL/OK Bridging

//OK
Freenect::Freenect freenect;
MyFreenectDevice* device;
MyFreenectDevice* devicetwo;
freenect_video_format requested_format(FREENECT_VIDEO_RGB);
double freenect_angle(0);
int got_frames(0),window(0);
int g_argc;
char **g_argv;
int user_data = 0;


//PCL
pcl::PointCloud<pcl::PointXYZRGB>::Ptr cloud (new pcl::PointCloud<pcl::PointXYZRGB>);
pcl::PointCloud<pcl::PointXYZRGB>::Ptr cloud2 (new pcl::PointCloud<pcl::PointXYZRGB>);
pcl::PointCloud<pcl::PointXYZRGB>::Ptr cloud1buf (new
pcl::PointCloud<pcl::PointXYZRGB>);
pcl::PointCloud<pcl::PointXYZRGB>::Ptr cloud2buf (new
pcl::PointCloud<pcl::PointXYZRGB>);

unsigned int cloud_id = 0;


///Keyboard Event Tracking
void keyboardEventOccurred (const pcl::visualization::KeyboardEvent &event,
                            void* viewer_void)
{
  boost::shared_ptr<pcl::visualization::PCLVisualizer> viewer =
*static_cast<boost::shared_ptr<pcl::visualization::PCLVisualizer> *> (viewer_void);
  if (event.getKeySym () == "c" && event.keyDown ())
  {
    std::cout << "c was pressed => capturing a pointcloud" << std::endl;
    std::string filename = "KinectCap";
    filename.append(boost::lexical_cast<std::string>(cloud_id));
```

```
        filename.append(".pcd");
        pcl::io::savePCDFileASCII (filename, *cloud);
        cloud_id++;
    }
}


// --------------
// -----Main-----
// --------------
int main (int argc, char** argv)
{
    Mat camera1Matrix, dist1Coeffs, camera2Matrix, dist2Coeffs;
    Mat R, T;
    vector<string> args;
    // copy program arguments into vector
    if (argc > 1) {
        for (int i=1;i<argc;i++)
        args.push_back(argv[i]);
    //LOAD THINGS!
    FileStorage fs(args[0], FileStorage::READ);
    fs["camera1Matrix"] >> camera1Matrix;
    fs["dist1Coeffs"] >> dist1Coeffs;
    fs["camera2Matrix"] >> camera2Matrix;
    fs["dist2Coeffs"] >> dist2Coeffs;
    fs["R"] >> R;
    fs["T"] >> T;
    fs.release();
    } else {
    printf("Wrong file to calib!");
    return 0;
    }


    //Create the Goal Transform for PCL
    Eigen::Vector3f PCTrans;
    Eigen::Quaternionf PCRot;
    Eigen::Matrix3f eRot;
    cv2eigen(R,eRot);
    PCRot = Eigen::Quaternionf(eRot);
    cv2eigen(T,PCTrans);
    PCTrans*=1000; //meters to mm


    //More Kinect Setup
    static std::vector<uint16_t> mdepth(640*480);
    static std::vector<uint8_t> mrgb(640*480*4);
    static std::vector<uint16_t> tdepth(640*480);
    static std::vector<uint8_t> trgb(640*480*4);

    // Fill in the cloud data
    cloud->width    = 640;
```

```
cloud->height   = 480;
cloud->is_dense = false;
cloud->points.resize (cloud->width * cloud->height);

// Fill in the cloud data
cloud2->width   = 640;
cloud2->height  = 480;
cloud2->is_dense = false;
cloud2->points.resize (cloud2->width * cloud2->height);

// Create and setup the viewer
printf("Create the viewer.\n");
boost::shared_ptr<pcl::visualization::PCLVisualizer> viewer (new
pcl::visualization::PCLVisualizer ("3D Viewer"));
viewer->registerKeyboardCallback (keyboardEventOccurred, (void*)&viewer);
viewer->setBackgroundColor (0, 0, 0);
viewer->addPointCloud<pcl::PointXYZRGB> (cloud, "Kinect Cloud");
viewer->addPointCloud<pcl::PointXYZRGB> (cloud2, "Second Cloud");
viewer->setPointCloudRenderingProperties
(pcl::visualization::PCL_VISUALIZER_POINT_SIZE, 1, "Kinect Cloud");
viewer->setPointCloudRenderingProperties
(pcl::visualization::PCL_VISUALIZER_POINT_SIZE, 1, "Second Cloud");
viewer->addCoordinateSystem (1.0);
viewer->initCameraParameters ();

printf("Create the devices.\n");
device = &freenect.createDevice<MyFreenectDevice>(0);
devicetwo = &freenect.createDevice<MyFreenectDevice>(1);
device->startVideo();
device->startDepth();
boost::this_thread::sleep (boost::posix_time::seconds (1));
devicetwo->startVideo();
devicetwo->startDepth();
boost::this_thread::sleep (boost::posix_time::seconds (1));
//Grab until clean returns
int DepthCount = 0;
while (DepthCount == 0) {
        device->updateState();
        device->getDepth(mdepth);
        device->getRGB(mrgb);
        for (size_t i = 0;i < 480*640;i++)
                DepthCount+=mdepth[i];
}

//-------------------
// -----Main loop-----
//-------------------
double x = NULL;
double y = NULL;
double tx = NULL;
double ty = NULL;
int iRealDepth = 0;
```

```
        int iTDepth = 0;

        device->setVideoFormat(requested_format);
        devicetwo->setVideoFormat(requested_format);
        printf("Start the main loop.\n");
        float percentCoverage = 90.0;
        while (!viewer->wasStopped ())
        {
                device->updateState();
                device->getDepth(mdepth);
                device->getRGB(mrgb);

                devicetwo->updateState();
                devicetwo->getDepth(tdepth);
                devicetwo->getRGB(trgb);

            size_t i = 0;
        size_t cinput = 0;
        int iMZeros = 0;
        int iTZeros = 0;
        for (size_t v=0 ; v<480 ; v++)
        {
                int mZeros = 0;
                int tZeros = 0;
                for ( size_t u=0 ; u<640 ; u++, i++)
                {
                        iRealDepth = mdepth[i];
                        if (iRealDepth == 0) mZeros++;
                        iTDepth = tdepth[i];
                        if (iTDepth == 0) tZeros++;
                                freenect_camera_to_world(device->getDevice(), u, v,
iRealDepth, &x, &y);
                                freenect_camera_to_world(devicetwo->getDevice(), u, v,
iTDepth, &tx, &ty);

                                cloud->points[i].x  = x;//1000.0;
                                cloud->points[i].y  = y;//1000.0;
                        cloud->points[i].z = iRealDepth;//1000.0;
                cloud->points[i].r = mrgb[i*3];
                        cloud->points[i].g = mrgb[(i*3)+1];
                cloud->points[i].b = mrgb[(i*3)+2];

                                cloud2->points[i].x  = tx;//1000.0;
                                cloud2->points[i].y  = ty;//1000.0;
                        cloud2->points[i].z = iTDepth;//1000.0;
                cloud2->points[i].r = trgb[i*3];
                        cloud2->points[i].g = trgb[(i*3)+1];
                cloud2->points[i].b = trgb[(i*3)+2];

                }
                //printf("mZeros = %d for column %d\n", mZeros, v);
                if (mZeros == 640) iMZeros++;
```

```
            if (tZeros == 640) iTZeros++;
            }

            //printf("iMZeros = %d for image.  iTZeros = %d for image.\n", iMZeros,
iTZeros);

            pcl::transformPointCloud (*cloud, *cloud, PCTrans, PCRot);

     //Only update the cloud if the other IR illuminator is 90% or better covered up.
     if (iTZeros/480 >= percentCoverage/100.0) {
            printf("Updating Cloud 1.\n");
            viewer->updatePointCloud (cloud, "Kinect Cloud");
     }
     if (iMZeros/480 >= percentCoverage/100.0) {
            printf("Updating Cloud 2.\n");
            viewer->updatePointCloud (cloud2, "Second Cloud");
     }

            viewer->spinOnce ();
     }
     device->stopVideo();
     device->stopDepth();
     devicetwo->stopVideo();
     devicetwo->stopDepth();
     return 0;
}
```

For this code, be sure to load a `calib.yaml` file (and if you didn't run OKStereo, do it now). You can see the results in Figures 10-11 and 10-12, the first being a cloud 1 update (Kinect 2's IR illuminator is covered) and the second a cloud 2 update (Kinect 1's IR illuminator is covered). Note the lack of interference in either figure. Figure 10-13 shows the combined point clouds from the top.

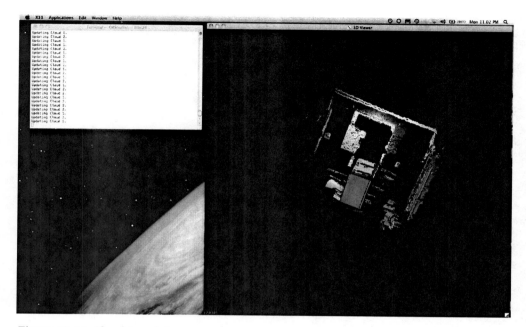

Figure 10-11. Cloud 1 update

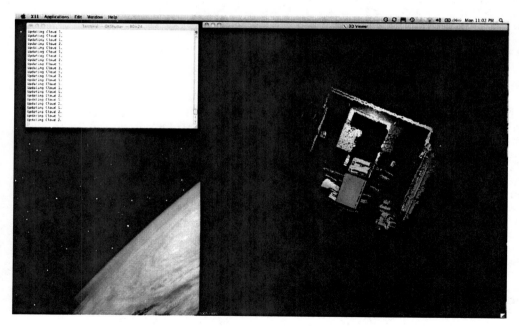

Figure 10-12. Cloud 2 update

Figure 10-13. Combined clouds from overhead

Calibration

Calibration between two Kinects is the establishment of a conversion between the world frame and each camera frame. A frame is the coordinate system that defines an object. A camera frame is the coordinate system for just that camera, or in this case, just that Kinect. For example, if we consider the center of the RGB camera 0,0,0 (origin) of the Kinect frame, the IR camera would be about 25 mm,0,0 or 25 mm in the positive x direction in the Kinect frame. The nice thing about defining frames is that you can set the frame origin wherever is most convenient for you. This is extremely handy when doing multiple systems because you have less calibration and conversion between frames.

The world frame is just what you expect—the coordinate system for the world (or in this case, our lab area). When we do a calibration, we center the world frame with one of our Kinect frames, then calibrate the other camera to this shared frame. This way we can save ourselves some work. You could also calibrate the target in the world frame, then fit the two Kinect frames to that frame.

To calibrate our Kinects together, we're going to perform a calibration for each individual Kinect (like in Chapter 3) and then use those calibrations to handle a calibration between the two Kinects. This second, stereo calibration is based on both Kinects imaging a known target with a known distribution of corners. Each frame that has a detected checkerboard for both cameras is stored as pixel points in an array. Those pixel points are then used to calculate the translation and rotation necessary to move one camera's frame into the other camera's frame. As our Kinects both have the same offset between the IR and RGB camera, rotating the frame works for both the depth and RGB image. The code in Listing 10-2 creates a `calib.yaml` file that you can import via the command line to short circuit the calibration procedure. You'll also need the checkerboard from Chapter 3.

First, you need to calibrate each camera individually. The live video output from the Kinect is used as the calibration window to show you how you're doing. Get the target as close as possible, while still having the points show up on the checkerboard (you'll see colored lines streaked across it). Move the checkerboard from side to side and make sure to reach into the corners and on all edges. If you get a value much greater than 1 for your error, try again. Get closer, get more light, or get closer to the edges. Repeat this for the second Kinect.

The next part is a little tricky. Move your checkerboard into a position that both Kinects can see. The counter in the terminal window will increase as you gather data. Gather 100 frames of stereo locked data, and you're home free.

▓ **Note** There is bug in the OpenCV Eigen code! As of this writing, the code for cv2eigen is broken. Mevatron on stackoverflow contributed a replacement file, which you can get it from this book's web site. It is included below for completeness as Listing 10-3. Be sure to reMake your OpenCV build with this file before attempting to run the code in Listing 10-2.

Listing 10-2. OKStereo.cpp

```
/*
 * This file is part of the OpenKinect Project. http://www.openkinect.org
 *
 * Copyright (c) 2010 individual OpenKinect contributors. See the CONTRIB file
 * for details.
 *
 * This code is licensed to you under the terms of the Apache License, version
 * 2.0, or, at your option, the terms of the GNU General Public License,
 * version 2.0. See the APACHE20 and GPL2 files for the text of the licenses,
 * or the following URLs:
 * http://www.apache.org/licenses/LICENSE-2.0
 * http://www.gnu.org/licenses/gpl-2.0.txt
 *
 * If you redistribute this file in source form, modified or unmodified, you
 * may:
 *   1) Leave this header intact and distribute it under the same terms,
 *      accompanying it with the APACHE20 and GPL20 files, or
 *   2) Delete the Apache 2.0 clause and accompany it with the GPL2 file, or
 *   3) Delete the GPL v2 clause and accompany it with the APACHE20 file
 * In all cases you must keep the copyright notice intact and include a copy
 * of the CONTRIB file.
 *
 * Binary distributions must follow the binary distribution requirements of
 * either License.
 */

#include <iostream>
```

```cpp
#include <libfreenect.hpp>
#include <pthread.h>
#include <stdio.h>
#include <string.h>
#include <cmath>
#include <vector>
#include <ctime>
#include <boost/thread/thread.hpp>
#include <algorithm>
#include <iterator>
#include <stdio.h>
#include <stdlib.h>
#include <ctype.h>
//OPENCV Includes
#include "opencv2/core/core.hpp"
#include "opencv2/imgproc/imgproc.hpp"
#include "opencv2/calib3d/calib3d.hpp"
#include "opencv2/highgui/highgui.hpp"
#include "opencv2/core/eigen.hpp"
//PCL Includes
#include "pcl/common/common_headers.h"
#include "pcl/common/eigen.h"
#include "pcl/common/transforms.h"
#include "pcl/features/normal_3d.h"
#include "pcl/io/pcd_io.h"
#include "pcl/visualization/pcl_visualizer.h"
#include "pcl/console/parse.h"
#include "pcl/point_types.h"
#include <pcl/kdtree/kdtree_flann.h>
#include <pcl/surface/mls.h>
#include "boost/lexical_cast.hpp"
#include "pcl/filters/voxel_grid.h"
#include "pcl/octree/octree.h"

using namespace cv;
using namespace std;

///Mutex Class
class Mutex {
public:
        Mutex() {
                pthread_mutex_init( &m_mutex, NULL );
        }
        void lock() {
                pthread_mutex_lock( &m_mutex );
        }
        void unlock() {
                pthread_mutex_unlock( &m_mutex );
        }

        class ScopedLock
        {
```

```cpp
                Mutex & _mutex;
        public:
                ScopedLock(Mutex & mutex)
                        : _mutex(mutex)
                {
                        _mutex.lock();
                }
                ~ScopedLock()
                {
                        _mutex.unlock();
                }
        };
private:
        pthread_mutex_t m_mutex;
};


///Kinect Hardware Connection Class
/* thanks to Yoda---- from IRC */
class MyFreenectDevice : public Freenect::FreenectDevice {
public:
        MyFreenectDevice(freenect_context * _ctx, int _index)
                : Freenect::FreenectDevice(_ctx, _index),
depth(freenect_find_depth_mode(FREENECT_RESOLUTION_MEDIUM,
FREENECT_DEPTH_REGISTERED).bytes),m_buffer_video(freenect_find_video_mode(FREENECT_RESOLUTION_
MEDIUM, FREENECT_VIDEO_RGB).bytes), m_new_rgb_frame(false), m_new_depth_frame(false)
        {

        }
        //~MyFreenectDevice(){}
        // Do not call directly even in child
        void VideoCallback(void* _rgb, uint32_t timestamp) {
                Mutex::ScopedLock lock(m_rgb_mutex);
                uint8_t* rgb = static_cast<uint8_t*>(_rgb);
                std::copy(rgb, rgb+getVideoBufferSize(), m_buffer_video.begin());
                m_new_rgb_frame = true;
        };
        // Do not call directly even in child
        void DepthCallback(void* _depth, uint32_t timestamp) {
                Mutex::ScopedLock lock(m_depth_mutex);
                depth.clear();
                uint16_t* call_depth = static_cast<uint16_t*>(_depth);
                for (size_t i = 0; i < 640*480 ; i++) {
                        depth.push_back(call_depth[i]);
                }
                m_new_depth_frame = true;
        }
        bool getRGB(std::vector<uint8_t> &buffer) {
                //printf("Getting RGB!\n");
                Mutex::ScopedLock lock(m_rgb_mutex);
                if (!m_new_rgb_frame) {
                        //printf("No new RGB Frame.\n");
```

```
                                return false;
                        }
                        buffer.swap(m_buffer_video);
                        m_new_rgb_frame = false;
                        return true;
                }

                bool getDepth(std::vector<uint16_t> &buffer) {
                        Mutex::ScopedLock lock(m_depth_mutex);
                        if (!m_new_depth_frame)
                                return false;
                        buffer.swap(depth);
                        m_new_depth_frame = false;
                        return true;
                }

        private:
                std::vector<uint16_t> depth;
                std::vector<uint8_t> m_buffer_video;
                Mutex m_rgb_mutex;
                Mutex m_depth_mutex;
                bool m_new_rgb_frame;
                bool m_new_depth_frame;
};


///Start the PCL/OK Bridging

//OK
Freenect::Freenect freenect;
MyFreenectDevice* device;
MyFreenectDevice* devicetwo;
freenect_video_format requested_format(FREENECT_VIDEO_RGB);
double freenect_angle(0);
int got_frames(0),window(0);
int g_argc;
char **g_argv;
int user_data = 0;

//PCL
pcl::PointCloud<pcl::PointXYZRGB>::Ptr cloud (new pcl::PointCloud<pcl::PointXYZRGB>);
pcl::PointCloud<pcl::PointXYZRGB>::Ptr cloud2 (new pcl::PointCloud<pcl::PointXYZRGB>);
pcl::PointCloud<pcl::PointXYZRGB>::Ptr bgcloud (new pcl::PointCloud<pcl::PointXYZRGB>);
pcl::PointCloud<pcl::PointXYZRGB>::Ptr voxcloud (new pcl::PointCloud<pcl::PointXYZRGB>);
float resolution = 50.0;
// Instantiate octree-based point cloud change detection class
pcl::octree::OctreePointCloudChangeDetector<pcl::PointXYZRGB> octree (resolution);

bool BackgroundSub = false;
bool hasBackground = false;
bool Voxelize = false;
unsigned int voxelsize = 10; //in mm
```

```
unsigned int cloud_id = 0;


//OpenCV
Mat mGray;
Size boardSize(10,7); //interior number of corners
Size imageSize;
float squareSize = 0.023; //23 mm
Mat camera1Matrix, dist1Coeffs, camera2Matrix, dist2Coeffs;
Mat R, T, E, F;
vector<vector<Point2f> > image1Points;
vector<vector<Point2f> > image2Points;
vector<Point2f> pointbuf, pointbuf2;
float aspectRatio = 1.0f;
vector<Mat> rvecs, tvecs;
vector<float> reprojErrs;
Mat map1, map2;
Mat mCalib;

static void calcChessboardCorners(Size boardSize, float squareSize, vector<Point3f>& corners)
{
    corners.resize(0);
    for( int i = 0; i < boardSize.height; i++ )
        for( int j = 0; j < boardSize.width; j++ )
            corners.push_back(Point3f(float(j*squareSize), float(i*squareSize), 0));
}

static double computeReprojectionErrors(
        const vector<vector<Point3f> >& objectPoints,
        const vector<vector<Point2f> >& imagePoints,
        const vector<Mat>& rvecs, const vector<Mat>& tvecs,
        const Mat& cameraMatrix, const Mat& distCoeffs,
        vector<float>& perViewErrors )
{
    vector<Point2f> imagePoints2;
    int i, totalPoints = 0;
    double totalErr = 0, err;
    perViewErrors.resize(objectPoints.size());

    for( i = 0; i < (int)objectPoints.size(); i++ )
    {
        projectPoints(Mat(objectPoints[i]), rvecs[i], tvecs[i],
                    cameraMatrix, distCoeffs, imagePoints2);
        err = norm(Mat(imagePoints[i]), Mat(imagePoints2), CV_L2);
        int n = (int)objectPoints[i].size();
        perViewErrors[i] = (float)std::sqrt(err*err/n);
        totalErr += err*err;
        totalPoints += n;
    }

    return std::sqrt(totalErr/totalPoints);
}
```

```
static bool runCalibration( vector<vector<Point2f> > imagePoints,
                    Size imageSize, Size boardSize,
                    float squareSize, float aspectRatio,
                    Mat& cameraMatrix, Mat& distCoeffs,
                    vector<Mat>& rvecs, vector<Mat>& tvecs,
                    vector<float>& reprojErrs,
                    double& totalAvgErr)
{
    cameraMatrix = Mat::eye(3, 3, CV_64F);

    distCoeffs = Mat::zeros(8, 1, CV_64F);

    vector<vector<Point3f> > objectPoints(1);
    calcChessboardCorners(boardSize, squareSize, objectPoints[0]);
    objectPoints.resize(imagePoints.size(),objectPoints[0]);

    double rms = calibrateCamera(objectPoints, imagePoints, imageSize, cameraMatrix,
                distCoeffs, rvecs, tvecs, CV_CALIB_FIX_K4|CV_CALIB_FIX_K5);
                ///*|CV_CALIB_FIX_K3*/|CV_CALIB_FIX_K4|CV_CALIB_FIX_K5);
    printf("RMS error reported by calibrateCamera: %g\n", rms);

    bool ok = checkRange(cameraMatrix) && checkRange(distCoeffs);

    totalAvgErr = computeReprojectionErrors(objectPoints, imagePoints, rvecs, tvecs,
cameraMatrix, distCoeffs, reprojErrs);

    return ok;
}

static bool runStereo ( vector<vector<Point2f> > image1Pt,
                                    vector<vector<Point2f> > image2Pt, Size imageSize,
Size boardSize, float squareSize, float aspectRatio, Mat c1Matrix, Mat c2Matrix, Mat d1Coeffs,
Mat d2Coeffs, Mat& R, Mat& T, Mat& E, Mat& F, double& totalAvgErr)
{

        vector<vector<Point3f> > objectPoints(1);
        calcChessboardCorners(boardSize, squareSize, objectPoints[0]);
        objectPoints.resize(image1Pt.size(),objectPoints[0]);

    double rms = stereoCalibrate(objectPoints, image1Pt, image2Pt,
                c1Matrix, d1Coeffs,
                c2Matrix, d2Coeffs,
                imageSize, R, T, E, F,
                TermCriteria(CV_TERMCRIT_ITER+CV_TERMCRIT_EPS, 100, 1e-5),
                CV_CALIB_FIX_INTRINSIC +
                CV_CALIB_FIX_ASPECT_RATIO +
                CV_CALIB_ZERO_TANGENT_DIST +
                CV_CALIB_SAME_FOCAL_LENGTH);
    cout << "Stereo Done with RMS Error=" << rms << endl;

    // CALIBRATION QUALITY CHECK
```

```
            // because the output fundamental matrix implicitly
            // includes all the output information,
            // we can check the quality of calibration using the
            // epipolar geometry constraint: m2^t*F*m1=0
    double err = 0;
    int npoints = 0;
    vector<Vec3f> lines[2];
    for(int i = 0; i < image1Pt.size(); i++ )
    {
        int npt = (int)image1Pt[i].size();
        Mat imgpt[2];
        imgpt[0] = Mat(image1Pt[i]);
            undistortPoints(imgpt[0], imgpt[0], c1Matrix, d1Coeffs, Mat(), c1Matrix);
        computeCorrespondEpilines(imgpt[0], 1, F, lines[0]);
        imgpt[1] = Mat(image2Pt[i]);
            undistortPoints(imgpt[1], imgpt[1], c2Matrix, d2Coeffs, Mat(), c2Matrix);
        computeCorrespondEpilines(imgpt[1], 2, F, lines[1]);

        for(int j = 0; j < npt; j++ )
        {
            double errij = fabs(image1Pt[i][j].x*lines[1][j][0] +
                            image1Pt[i][j].y*lines[1][j][1] + lines[1][j][2]) +
                        fabs(image2Pt[i][j].x*lines[0][j][0] +
                            image2Pt[i][j].y*lines[0][j][1] + lines[0][j][2]);
            err += errij;
        }
        npoints += npt;
    }
    totalAvgErr = err/npoints;

    return true;

}


///Keyboard Event Tracking
void keyboardEventOccurred (const pcl::visualization::KeyboardEvent &event,
                            void* viewer_void)
{
  boost::shared_ptr<pcl::visualization::PCLVisualizer> viewer =
*static_cast<boost::shared_ptr<pcl::visualization::PCLVisualizer> *> (viewer_void);
  if (event.getKeySym () == "c" && event.keyDown ())
  {
    std::cout << "c was pressed => capturing a pointcloud" << std::endl;
    std::string filename = "KinectCap";
    filename.append(boost::lexical_cast<std::string>(cloud_id));
    filename.append(".pcd");
    pcl::io::savePCDFileASCII (filename, *cloud);
    cloud_id++;
  }
}
```

```cpp
// --------------
// -----Main-----
// --------------
int main (int argc, char** argv)
{
        int State = 0; //0 = Calib 1, 1 = Calib 2, 2 = Calib Stereo, 3 = PCL convert
        bool load = false;
        // create an empty vector of strings
    vector<string> args;
    // copy program arguments into vector
    if (argc > 1) {
            for (int i=1;i<argc;i++)
            args.push_back(argv[i]);
        //LOAD THINGS!
        load = true;
        FileStorage fs(args[0], FileStorage::READ);
        fs["camera1Matrix"] >> camera1Matrix;
        fs["dist1Coeffs"] >> dist1Coeffs;
        fs["camera2Matrix"] >> camera2Matrix;
        fs["dist2Coeffs"] >> dist2Coeffs;
        fs["R"] >> R;
        fs["T"] >> T;
        fs.release();
        State = 3;
    }


        //More Kinect Setup
        static std::vector<uint16_t> kdepth(640*480);
        static std::vector<uint8_t> krgb(640*480*4);
        static std::vector<uint16_t> tdepth(640*480);
        static std::vector<uint8_t> trgb(640*480*4);

        // Create and setup OpenCV
    Mat mRGB (480, 640, CV_8UC3);
    Mat mDepth (480, 640, CV_16UC1);
    Mat tRGB (480, 640, CV_8UC3);
    Mat tDepth (480, 640, CV_16UC1);
    imageSize = mRGB.size();

        if (!load)
            cvNamedWindow("Image", CV_WINDOW_AUTOSIZE);

    // Fill in the cloud data
        cloud->width    = 640;
        cloud->height   = 480;
        cloud->is_dense = false;
        cloud->points.resize (cloud->width * cloud->height);

        // Fill in the cloud data
        cloud2->width   = 640;
        cloud2->height  = 480;
```

```
            cloud2->is_dense = false;
            cloud2->points.resize (cloud2->width * cloud2->height);
            //Create the Goal Transform for PCL
            Eigen::Vector3f PCTrans;
            Eigen::Quaternionf PCRot;

            // Create the viewer
            boost::shared_ptr <pcl::visualization::PCLVisualizer> viewer;

            //Loaded?
            if (load) {
                    Eigen::Matrix3f eRot;
                    cv2eigen(R,eRot);
                    PCRot = Eigen::Quaternionf(eRot);
                    cv2eigen(T,PCTrans);
                    PCTrans*=1000; //meters to mm

                    //Open PCL section
                    viewer = boost::shared_ptr<pcl::visualization::PCLVisualizer> (new
        pcl::visualization::PCLVisualizer ("3D Viewer"));
                    boost::this_thread::sleep (boost::posix_time::seconds (1));

                    viewer->registerKeyboardCallback (keyboardEventOccurred, (void*)&viewer);
                    viewer->setBackgroundColor (255, 255, 255);
                    viewer->addPointCloud<pcl::PointXYZRGB> (cloud, "Kinect Cloud");
                    viewer->addPointCloud<pcl::PointXYZRGB> (cloud2, "Second Cloud");
                    viewer->setPointCloudRenderingProperties
        (pcl::visualization::PCL_VISUALIZER_POINT_SIZE, 1, "Kinect Cloud");
                    viewer->setPointCloudRenderingProperties
        (pcl::visualization::PCL_VISUALIZER_POINT_SIZE, 1, "Second Cloud");
                    viewer->addCoordinateSystem (1.0);
                    viewer->initCameraParameters ();
                    printf("Viewer Built!  Displaying 3D Point Clouds\n");
            }

            printf("Create the devices.\n");
            device = &freenect.createDevice<MyFreenectDevice>(0);
            devicetwo = &freenect.createDevice<MyFreenectDevice>(1);
            device->startVideo();
            device->startDepth();
            boost::this_thread::sleep (boost::posix_time::seconds (1));
            devicetwo->startVideo();
            devicetwo->startDepth();
            boost::this_thread::sleep (boost::posix_time::seconds (1));

            //Grab until clean returns
            int DepthCount = 0;
            while (DepthCount == 0) {
                    device->updateState();
                    device->getDepth(kdepth);
                    device->getRGB(krgb);
                    for (size_t i = 0;i < 480*640;i++)
```

```
                        DepthCount+=kdepth[i];
        }

        //--------------------
        // -----Main loop-----
        //--------------------
        double x = NULL;
        double y = NULL;
        double tx = NULL;
        double ty = NULL;
        int iRealDepth = 0;
        int iTDepth = 0;

        device->setVideoFormat(requested_format);
        devicetwo->setVideoFormat(requested_format);
        printf("Start the main loop.\n");


    while (1) {
                device->updateState();
                device->getDepth(kdepth);
                device->getRGB(krgb);

                devicetwo->updateState();
                devicetwo->getDepth(tdepth);
                devicetwo->getRGB(trgb);

            size_t i = 0;
              size_t cinput = 0;


            for (size_t v=0 ; v<480 ; v++)
        {
                        uint8_t* rowRPtr = mRGB.ptr<uint8_t>(v);
                        uint16_t* rowDPtr = mDepth.ptr<uint16_t>(v);
                        uint8_t* rowTRPtr = tRGB.ptr<uint8_t>(v);
                        uint16_t* rowTDPtr = tDepth.ptr<uint16_t>(v);
                        cinput = 0;
                    for ( size_t u=0 ; u<640 ; u++, i++, cinput++)
                    {
                        iRealDepth = kdepth[i];
                        iTDepth = tdepth[i];
                        freenect_camera_to_world(device->getDevice(), u, v, iRealDepth, &x,
&y);
                            freenect_camera_to_world(devicetwo->getDevice(), u, v,
iTDepth, &tx, &ty);
                        rowDPtr[u] = iRealDepth;
                        rowRPtr[(cinput*3)] = krgb[(i*3)+2];
                        rowRPtr[(cinput*3)+1] = krgb[(i*3)+1];
                        rowRPtr[(cinput*3)+2] = krgb[(i*3)];
                        rowTDPtr[u] = iTDepth;
                            rowTRPtr[(cinput*3)] = trgb[(i*3)+2];
```

```
                rowTRPtr[(cinput*3)+1] = trgb[(i*3)+1];
                       rowTRPtr[(cinput*3)+2] = trgb[(i*3)];

                       cloud->points[i].x  = x;//1000.0;
                       cloud->points[i].y  = y;//1000.0;
            cloud->points[i].z = iRealDepth;//1000.0;
                cloud->points[i].r = krgb[i*3];
            cloud->points[i].g = krgb[(i*3)+1];
                cloud->points[i].b = krgb[(i*3)+2];

                       cloud2->points[i].x  = tx;//1000.0;
                       cloud2->points[i].y  = ty;//1000.0;
            cloud2->points[i].z = iTDepth;//1000.0;
        cloud2->points[i].r = trgb[i*3];
           cloud2->points[i].g = trgb[(i*3)+1];
                cloud2->points[i].b = trgb[(i*3)+2];
        }
        }

        //printf("Displaying mRGB of depth %d\n", mRGB.depth());

        if (!load && State < 3) {
                if (State == 0) {
                        cvtColor( mRGB, mGray, CV_RGB2GRAY );
                        bool found = false;
                        //CALIB_CB_FAST_CHECK saves a lot of time on images
                        //that do not contain any chessboard corners
                        found = findChessboardCorners( mGray, boardSize, pointbuf,
                CV_CALIB_CB_ADAPTIVE_THRESH | CV_CALIB_CB_FAST_CHECK |
CV_CALIB_CB_NORMALIZE_IMAGE);

                        if(found)
                        {
                                cornerSubPix( mGray, pointbuf, Size(11,11),
                Size(-1,-1), TermCriteria( CV_TERMCRIT_EPS+CV_TERMCRIT_ITER, 30, 0.1
));

                image1Points.push_back(pointbuf);
                drawChessboardCorners( mRGB, boardSize, Mat(pointbuf), found );
                }

                imshow("Image", mRGB);
                }
                else if (State == 1) {
                        cvtColor( tRGB, mGray, CV_RGB2GRAY );
                        bool found = false;
                        //CALIB_CB_FAST_CHECK saves a lot of time on images
                        //that do not contain any chessboard corners
                        found = findChessboardCorners( mGray, boardSize, pointbuf,
                CV_CALIB_CB_ADAPTIVE_THRESH | CV_CALIB_CB_FAST_CHECK |
CV_CALIB_CB_NORMALIZE_IMAGE);

                        if(found)
                        {
```

```
                                        cornerSubPix( mGray, pointbuf, Size(11,11),
                            Size(-1,-1), TermCriteria( CV_TERMCRIT_EPS+CV_TERMCRIT_ITER, 30, 0.1
));

                            image2Points.push_back(pointbuf);
                            drawChessboardCorners( tRGB, boardSize, Mat(pointbuf), found );
                    }

                            imshow("Image", tRGB);
                    }
                else if (State == 2) {
                    //Stereo Calibration
                    cvtColor( mRGB, mGray, CV_RGB2GRAY );
                            bool found1 = false;
                            //CALIB_CB_FAST_CHECK saves a lot of time on images
                            //that do not contain any chessboard corners
                            found1 = findChessboardCorners( mGray, boardSize, pointbuf,
                    CV_CALIB_CB_ADAPTIVE_THRESH | CV_CALIB_CB_FAST_CHECK |
CV_CALIB_CB_NORMALIZE_IMAGE);

                                    if(found1)
                                    {
                                            cornerSubPix( mGray, pointbuf, Size(11,11),
                            Size(-1,-1), TermCriteria( CV_TERMCRIT_EPS+CV_TERMCRIT_ITER, 30, 0.1
));

                                    drawChessboardCorners( mRGB, boardSize, Mat(pointbuf), found1);
                    }
                imshow("Image", mRGB);
                            cvtColor( tRGB, mGray, CV_RGB2GRAY );
                            bool found2 = false;
                            //CALIB_CB_FAST_CHECK saves a lot of time on images
                            //that do not contain any chessboard corners
                            found2 = findChessboardCorners( mGray, boardSize, pointbuf2,
                    CV_CALIB_CB_ADAPTIVE_THRESH | CV_CALIB_CB_FAST_CHECK |
CV_CALIB_CB_NORMALIZE_IMAGE);

                                    if(found2)
                                    {
                                            cornerSubPix( mGray, pointbuf2, Size(11,11),
                            Size(-1,-1), TermCriteria( CV_TERMCRIT_EPS+CV_TERMCRIT_ITER, 30, 0.1
));

                                    drawChessboardCorners( tRGB, boardSize, Mat(pointbuf2), found2 );
                                    imshow("Image", tRGB);
                    }

                if (found1 && found2)
                {
                        image1Points.push_back(pointbuf);
                        image2Points.push_back(pointbuf2);
                        printf("%d\n", image1Points.size());
                }

                }
```

```
                if (State == 0 && image1Points.size() >= 100) {
                        cout << "Calculating Distortion and Camera Matrix for Image 1"
<< endl;

                        State = 1;
                        double totalAvgErr = 0;

                        bool ok = runCalibration(image1Points, imageSize, boardSize,
squareSize, aspectRatio, camera1Matrix, dist1Coeffs, rvecs, tvecs, reprojErrs, totalAvgErr);
                                printf("%s. avg reprojection error = %.2f\n", ok ? "Calibration
succeeded" : "Calibration failed", totalAvgErr);

                        cout << "Camera Matrix: " << camera1Matrix << endl;
                    cout << "Dist Coeffs: " << dist1Coeffs << endl;
                }

                if (State == 1 && image2Points.size() >= 100) {
                        cout << "Calculating Distortion and Camera Matrix for Image 2"
<< endl;

                        State = 2;
                        double totalAvgErr = 0;

                        bool ok = runCalibration(image2Points, imageSize, boardSize,
squareSize, aspectRatio, camera2Matrix, dist2Coeffs, rvecs, tvecs, reprojErrs, totalAvgErr);
                                printf("%s. avg reprojection error = %.2f\n", ok ? "Calibration
succeeded" : "Calibration failed", totalAvgErr);

                        cout << "Camera Matrix: " << camera2Matrix << endl;
                    cout << "Dist Coeffs: " << dist2Coeffs << endl;
                    image1Points.clear();
                    image2Points.clear();
                }

                if (State == 2 && image1Points.size() >= 100 && image2Points.size() >=
100)
                {
                    State = 3;
                    double totalAvgErr = 0;


                    bool ok = runStereo ( image1Points,
                                image2Points, imageSize, boardSize, squareSize, aspectRatio,
camera1Matrix, camera2Matrix, dist1Coeffs, dist2Coeffs, R, T, E, F, totalAvgErr);
                    cout << "Stereo Avg Repro Err" << totalAvgErr << endl;
                    cout << "Rotation: " << R << endl;
                    cout << "Translation: " << T << endl;

                    Eigen::Matrix3f eRot;
                    cv2eigen(R,eRot);
                    PCRot = Eigen::Quaternionf(eRot);
                    cv2eigen(T,PCTrans);
                    PCTrans*=1000; //meters to mm
```

```
                        //Store the Data from this calib
                        printf("Writing calib out to calib.yaml.");
                        FileStorage fs("calib.yaml", FileStorage::WRITE);
                        fs << "camera1Matrix" << camera1Matrix;
                        fs << "dist1Coeffs"  << dist1Coeffs;
                        fs << "camera2Matrix" << camera2Matrix;
                        fs << "dist2Coeffs" << dist2Coeffs;
                        fs << "R" << R;
                        fs << "T" << T;
                        fs.release();

                        //Close out OpenCV section
                        destroyAllWindows();
                                boost::this_thread::sleep (boost::posix_time::seconds (1));


                                //Open PCL section
                                viewer = boost::shared_ptr<pcl::visualization::PCLVisualizer>
(new pcl::visualization::PCLVisualizer ("3D Viewer"));
                                boost::this_thread::sleep (boost::posix_time::seconds (1));

                                viewer->registerKeyboardCallback (keyboardEventOccurred,
(void*)&viewer);
                                viewer->setBackgroundColor (255, 255, 255);
                                viewer->addPointCloud<pcl::PointXYZRGB> (cloud, "Kinect
Cloud");
                                viewer->addPointCloud<pcl::PointXYZRGB> (cloud2, "Second
Cloud");
                                viewer->setPointCloudRenderingProperties
(pcl::visualization::PCL_VISUALIZER_POINT_SIZE, 1, "Kinect Cloud");
                                viewer->setPointCloudRenderingProperties
(pcl::visualization::PCL_VISUALIZER_POINT_SIZE, 1, "Second Cloud");
                                viewer->addCoordinateSystem (1.0);
                                viewer->initCameraParameters ();
                                printf("Viewer Built!  Displaying 3D Point Clouds\n");
                                continue;
                }

                        cvWaitKey(66);
                }
                else
                {
                        if (!viewer->wasStopped ())
                        {
                                pcl::transformPointCloud (*cloud, *cloud, PCTrans, PCRot);
                        viewer->updatePointCloud (cloud, "Kinect Cloud");
                        viewer->updatePointCloud (cloud2, "Second Cloud");
                        viewer->spinOnce ();
                }
                else
                {
```

```
                            break;
                    }
                }
            }
            mRGB.release();
            mDepth.release();
            tRGB.release();
            tDepth.release();
            device->stopVideo();
            device->stopDepth();
            devicetwo->stopVideo();
            devicetwo->stopDepth();
            return 0;

}
```

Listing 10-3. eigen.hpp

```
#ifndef __OPENCV_CORE_EIGEN_HPP__
#define __OPENCV_CORE_EIGEN_HPP__

#ifdef __cplusplus

#include "opencv/cxcore.h"
#include <eigen3/Eigen/Dense>

namespace cv
{

template<typename _Tp, int _rows, int _cols, int _options, int _maxRows, int _maxCols>
void eigen2cv( const Eigen::Matrix<_Tp, _rows, _cols, _options, _maxRows, _maxCols>& src, Mat&
dst )
{
    if( !(src.Flags & Eigen::RowMajorBit) )
    {
        Mat _src(src.cols(), src.rows(), DataType<_Tp>::type,
                (void*)src.data(), src.stride()*sizeof(_Tp));
        transpose(_src, dst);
    }
    else
    {
        Mat _src(src.rows(), src.cols(), DataType<_Tp>::type,
                (void*)src.data(), src.stride()*sizeof(_Tp));
        _src.copyTo(dst);
    }
}

template<typename _Tp, int _rows, int _cols, int _options, int _maxRows, int _maxCols>
void cv2eigen( const Mat& src,
            Eigen::Matrix<_Tp, _rows, _cols, _options, _maxRows, _maxCols>& dst )
{
```

243

```
    CV_DbgAssert(src.rows == _rows && src.cols == _cols);
    if( !(dst.Flags & Eigen::RowMajorBit) )
    {
        Mat _dst(src.cols, src.rows, DataType<_Tp>::type,
                 dst.data(), (size_t)(dst.stride()*sizeof(_Tp)));
        if( src.type() == _dst.type() )
            transpose(src, _dst);
        else if( src.cols == src.rows )
        {
            src.convertTo(_dst, _dst.type());
            transpose(_dst, _dst);
        }
        else
            Mat(src.t()).convertTo(_dst, _dst.type());
        CV_DbgAssert(_dst.data == (uchar*)dst.data());
    }
    else
    {
        Mat _dst(src.rows, src.cols, DataType<_Tp>::type,
                 dst.data(), (size_t)(dst.stride()*sizeof(_Tp)));
        src.convertTo(_dst, _dst.type());
        CV_DbgAssert(_dst.data == (uchar*)dst.data());
    }
}

template<typename _Tp>
void cv2eigen( const Mat& src,
              Eigen::Matrix<_Tp, Eigen::Dynamic, Eigen::Dynamic>& dst )
{
    dst.resize(src.rows, src.cols);
    if( !(dst.Flags & Eigen::RowMajorBit) )
    {
        Mat _dst(src.cols, src.rows, DataType<_Tp>::type,
            dst.data(), (size_t)(dst.stride()*sizeof(_Tp)));
        if( src.type() == _dst.type() )
            transpose(src, _dst);
        else if( src.cols == src.rows )
        {
            src.convertTo(_dst, _dst.type());
            transpose(_dst, _dst);
        }
        else
            Mat(src.t()).convertTo(_dst, _dst.type());
        CV_DbgAssert(_dst.data == (uchar*)dst.data());
    }
    else
    {
        Mat _dst(src.rows, src.cols, DataType<_Tp>::type,
                 dst.data(), (size_t)(dst.stride()*sizeof(_Tp)));
        src.convertTo(_dst, _dst.type());
        CV_DbgAssert(_dst.data == (uchar*)dst.data());
    }
```

```
}


template<typename _Tp>
void cv2eigen( const Mat& src,
                Eigen::Matrix<_Tp, Eigen::Dynamic, 1>& dst )
{
    CV_Assert(src.cols == 1);
    dst.resize(src.rows);

    if( !(dst.Flags & Eigen::RowMajorBit) )
    {
        Mat _dst(src.cols, src.rows, DataType<_Tp>::type,
                dst.data(), (size_t)(dst.stride()*sizeof(_Tp)));
        if( src.type() == _dst.type() )
            transpose(src, _dst);
        else
            Mat(src.t()).convertTo(_dst, _dst.type());
        CV_DbgAssert(_dst.data == (uchar*)dst.data());
    }
    else
    {
        Mat _dst(src.rows, src.cols, DataType<_Tp>::type,
                dst.data(), (size_t)(dst.stride()*sizeof(_Tp)));
        src.convertTo(_dst, _dst.type());
        CV_DbgAssert(_dst.data == (uchar*)dst.data());
    }
}


template<typename _Tp>
void cv2eigen( const Mat& src,
                Eigen::Matrix<_Tp, 1, Eigen::Dynamic>& dst )
{
    CV_Assert(src.rows == 1);
    dst.resize(src.cols);
    if( !(dst.Flags & Eigen::RowMajorBit) )
    {
        Mat _dst(src.cols, src.rows, DataType<_Tp>::type,
                dst.data(), (size_t)(dst.stride()*sizeof(_Tp)));
        if( src.type() == _dst.type() )
            transpose(src, _dst);
        else
            Mat(src.t()).convertTo(_dst, _dst.type());
        CV_DbgAssert(_dst.data == (uchar*)dst.data());
    }
    else
    {
        Mat _dst(src.rows, src.cols, DataType<_Tp>::type,
                dst.data(), (size_t)(dst.stride()*sizeof(_Tp)));
        src.convertTo(_dst, _dst.type());
        CV_DbgAssert(_dst.data == (uchar*)dst.data());
```

```
      }
  }

  }

#endif

#endif
```

Summary

In this chapter, you learned why you'd want to use multiple Kinects in a project. You looked at some of the limitations and issues when using multiple Kinects and how to overcome them. Then we showed you a mechanical shutter system and the code to use with it in order to eliminate interference. Finally, we showed you an on-the-fly calibration system for two Kinects.

Index

A

Ambient light, 12

B

Blurring algorithm, 68
Brightness thresholding algorithm
    float testApp::blur, 70
    kinect.update(), 69
    void testApp::draw(), 70
    void testApp::exit(), 70
    void testApp::setup(), 69
    void testApp::update(), 69

C, D, E, F

Computer vision
    image anatomy, 65
    image comparison, 74
        background subtraction, 76–80
        black and white image creation, 85–86
        double image, 85
        frame differencing, 80–84
        image storage, 87
        tolerance, 75–76
    image processing. *see* Image processing

G

Gesture recognition, 89
    definiton, 89
    multitouch detection
        assigning and tracking component IDs, 95–96
        camera image, background storing and subtracting, 92
        connected components algorithm, 93–95
        fingertip touching, 90
        image processing, 90–91
        infrared emitter and detector, 89
        Kinect's depth image, 101
        LCD display, 89
        minority report—style interface, 99–100
        motion, 97
        multitouch-capable devices, 96
        rotation, 97–99
        scale, 99
        shape, 101
        threshold filter, 90, 92–93

H

Hardware
    accelerometer, 23–24
    depth sensing, 11–13
    RGB camera. *see* RGB camera
    tilting head, 23–24
    volumetric sensing
        Arduino Sketch, 35, 36, 38
        binary distributions, 26
        block wiring, 37
        //BufferedAsync Setup, 32, 34
        CMakeLists.txt file, 34–35
        ///Keyboard Event Tracking, 29, 30
        ///Kinect hardware connection class, 27
        lit alarm light, 39
        ///Mutex Class, 26, 27
        //~MyFreenectDevice(), 27
        parts, 25
        //PCL, 29
        //Percentage Change, 30, 32
        relay wiring, 35, 36
        ///Start the PCL/OK Bridging, 28–29

I, J

Image processing
    brightest pixel tracking, 72–74
    brightness thresholding algorithm. *see*
        Brightness thresholding algorithm

Image processing (*cont.*)
  data simplifying, 66–67
  noise and blurring, 67–68
  situation contriving, 69
Infrared emitter and detector, 89

▓ K

Kinect
  drivers installation:. *see* OpenKinect driver
  hardware requirements, 1–2
  installation testing, 9

▓ L

Linux, 6–7

▓ M, N

Mac OS X, 8–9
Mesh Models:, 128
Multiple kinects
  calibration, 209
    calib.yaml file, 229
    camera frame, 228
    eigen.hpp, 243
    OKStereo.cpp, 229–243
    stereo calibration, 228
    world frame, 228
  depth shadows, occlusions, 208
  field of view, 207
  hardware requirements, 209
  interference, 209
    angle and distance, 213, 214
    box fan test, 217
    calib.yaml file, 226
    cloud 1 update, 227
    cloud 2 update, 227
    combined point clouds, 226
    hardware shutter system, 218
    holes, 209, 212
    IR camera, 218
    IR pattern, 209–211
    laser diodes, 217
    mechanical shutters, 217
    noise, 209
    OKShutter.cpp, 219–226
    scene, 213
    splotches, 209, 213
  single direction, 207

Multitouch detection, gesture recognition
  assigning and tracking component IDs, 95–96
  camera image, background storing and
    subtracting, 92
  connected components algorithm, 93–95
  fingertip touching, 90
  image processing, 90–91
  infrared emitter and detector, 89
  Kinect's depth image, 101
  LCD display, 89
  minority report—style interface, 99–100
  motion, 97
  multitouch-capable devices, 96
  rotation, 97–99
  scale, 99
  shape, 101
  threshold filter, 90, 92–93

▓ O

Object detection
  global descriptors
    CloudRecognizer Class, 201, 202
    database model, 202
    VFH descriptor computation, 202–204
  pose estimation, 204–205
Object modeling
  3-D
    camera space, 191
    cleaning and cropping, partial views, 195–
      196
    high-resolution models, 199
    Kinect pose estimation, 192–195
    marker-based scanner, 189–190
    Point clouds merging, 196–199
    support builiding, 191
  definition, 173
  single Kinect image. *see* Single Kinect image
OpenGL
  drawing points, 135
  initialization code, 133–135
OpenKinect driver
  Linux, 6–7
  Mac OS X, 8–9
  Red Hat/Fedora, 7
  Ubuntu, 7
  Windows, 2
    CMake preconfiguration, 5
    Git Commands, 3
    libfreenect, 3

Microsoft Visual Studio 2010 and MinGW, 5–6
updation, 3

▨ P, Q

Person tracking, 123–125
Point cloud library (PCL), 129
  OpenKinect
    binary distributions, 57
    // Create and setup the viewer, 60
    C++ file creation, 56
    —CMakeLists.txt—, 62
    ///Kinect Hardware Connection Class, 58
    ///Mutex Class, 58
    //~MyFreenectDevice(), 58
    //More Kinect Setup, 60
    ///Start the PCL/OK Bridging, 59
  Windows installation
    cmake-guifor FLANN, 49
    CMinPack, 49
    Linux, 52–53
    Mac OS X, 53–55
    Qhull, 51
    VTK installer, 51
Point clouds
  coloring
    depth to color reference frame, 131
    image plane, 132
  Depth Map, 130–131
  2-D registration
    affine transformation, 154
    matched features, 152
    transformation parameters, 153
    translation, 153, 154
  3-D data representation
    Mesh Models, 128
    rendering, 129
    scaling pixel count, 127, 128
    Voxels, 128
  3-D registration
    absolute orientation, 156–157
    rigid transformation, 155
    rotation computation, 155
  outliers, 157–159
  PCL creation, 139–142
  SLAM. See Simultaneous Localization and Mapping
  surface reconstruction
    normal estimation, 162

triangulation method, 162–163
  visualization
    with OpenGL, 133–135
    with PCL, 133
  wind application
    blue-red gradient, 136
    Freenect Thread Code, 137–139
    intensity field, 142
    is_frozen, 142
    Kinect depth image, 139–142
    libraries, 136–137
    main() function, 139
    OpenGL Code, 143–149
    radiohead's video, 136
    screenshot, 149
    show_visualizer(), 142
    structure of, 136
    TMyPoint, 142

▨ R

Random sample consensus (RANSAC), 174
Red Hat/Fedora, 7
RGB camera
  build/bin/calibrate_kinect_ir execution, 18
  calibration target, 13, 16–17
  Capture directory, 17–18
  Combined R|T matrix, 23
  kinect_calibration.yml file, 19–21
  Linux, 15
  Mac OS X, 16
  output image, 18, 19
  pinhole model, 21, 22
  rgb_distortion and depth_distortion, 21, 22
  rgb_intrinsics/depth_intrinsics, 21
  rgbd-viewer, 17
  Windows installation, 14–15

▨ S

Shape gestures, 101
Simultaneous localization and mapping (SLAM)
  advantages of, 160
  conventional camera, 159
  Kinect, 160–161
  real-time considerations, 161
  simple Kinect
    C++ classes, 164
    camera pose estimation, 170–172
    CTrackingSharedData class, 166

SLAM, simple kinect (*cont.*)
    main classes of, 164–165
    median feature computation, 168
    Point Map construction, 169–170
    screenshot, 164
    SURF, 167–168
Single Kinect image
    3-D Model
        extruder class, 180
        Mesh building, 187–188
        surface point cloud, 181, 185–187
        unseen Voxels, 183–185
        Voxelized representation, 181–183
    parametric model, 178–179
    tabletop object detector
        background removal, 176
        individual object clusters extraction, 177–178
        points lying, prism, 177
        sample scene, 174
        table plane extraction, 174–176
Software
    Kinect drivers
        Microsoft Kinect SDK, 41
        OpenKinect, 41
        OpenNI, 41
    OpenCV installation
        Linux, 52–53
        Mac OS X, 53–55
        Windows, 42–43
    point cloud library (PCL) installation
        // Create and setup the viewer, 60
        ///Kinect Hardware Connection Class, 58
        ///Mutex Class, 58
        ///Start the PCL/OK Bridging, 59
        //~MyFreenectDevice(), 58
        //More Kinect Setup, 60
        binary distributions, 57
        C++ file creation, 56
        —CMakeLists.txt—, 62
Structured light pattern, 12

T

Tabletop object detector
    background removal, 176
    individual object clusters extraction, 177–178
    points lying, prism, 177
    sample scene, 174
    table plane extraction, 174–176

Threshold filter, 92–93

U

Ubuntu, 7

V

Volumetric sensing
    OKFlower.cpp
        Arduino Sketch, 35, 36, 38
        binary distributions, 26
        block wiring, 37
        //BufferedAsync Setup, 32, 34
        CMakeLists.txt file, 34–35
        ///Keyboard Event Tracking, 29, 30
        ///Kinect hardware connection class, 27
        lit alarm light, 39
        ///Mutex Class, 26, 27
        //~MyFreenectDevice(), 27
        //PCL, 29
        //Percentage Change, 30, 32
        relay wiring, 35, 36
        ///Start the PCL/OK Bridging, 28–29
    parts, 25
Voxelization, 103
    clustering voxels, 122
        cluster_indices, 122
        2-D flood fill technique, 120
        EuclideanClusterExtraction, 121
        KdTree line, 122
        PCL, 121–122
        setClusterTolerance, 122
        setMinClusterSize and setMaxClusterSize, 122
    dataset, 104
    definition, 103–104
    manipulating voxels
        background cloud, 118
        background subtraction, 108–116
        drawing voxel boxes, 108
        foreground cloud, 117, 120
        full scene cloud, 117, 119
        function, background subtraction, 116–117
        getPointIndicesFromNewVoxels, 117
        leaf nodes, 107
    octrees, 105–107
    PCL, 105

tracking people and fitting rectangular prism, 122–125
Voxels, 128

▓ W, X, Y, Z

Wind application
    animation code, 142–143
    blue-red gradient, 136
    Freenect Thread Code, 137–139
    intensity field, 142
    is_frozen, 142
    Kinect depth image, 139–142
    libraries, 136–137
    main() function, 139
    OpenGL Code, 143–149
    screenshot, 149
    show_visualizer(), 142
    structure of, 136
    TMyPoint, 142

CPSIA information can be obtained at www.ICGtesting.com
Printed in the USA
LVOW111359040412

276141LV00003B/15/P